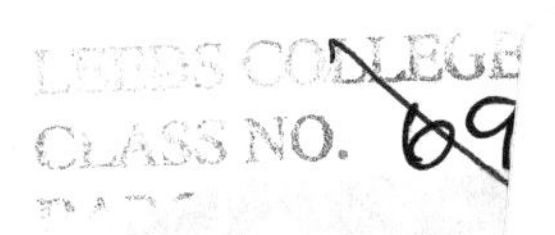

LEEDS COLLEGE OF BUILDING LIBRARY
CLASS NO. 69
I0996233

Macmillan Building and Surveying Series

Accounting and Finance for Building and Surveying A.R. Jennings
Advanced Building Measurement, second edition Ivor H. Seeley
Advanced Valuation Diane Butler and David Richmond
Applied Valuation, second edition Diane Butler
Asset Valuation Michael Rayner
Auctioning Real Property R.M. Courtenay Lord
Building Economics, fourth edition Ivor H. Seeley
Building Maintenance, second edition Ivor H. Seeley
Building Maintenance Technology Lee How Son and George C.S. Yuen
Building Procurement, second edition Alan E. Turner
Building Project Appraisal Keith Hutchinson
Building Quantities Explained, fifth edition Ivor H. Seeley
Building Services George Hassan
Building Surveys, Reports and Dilapidations Ivor H. Seeley
Building Technology, fifth edition Ivor H. Seeley
Civil Engineering Contract Administration and Control, second edition Ivor H. Seeley
Civil Engineering Quantities, fifth edition Ivor H. Seeley
Commercial Lease Renewals – A Practical Guide Philip Freedman and Eric F. Shapiro
Computers and Quantity Surveyors A.J. Smith
Conflicts in Construction – Avoiding, managing, resolving Jeff Whitfield
Constructability in Building and Engineering Projects A. Griffith and A.C. Sidwell
Construction Contract Claims Reg Thomas
Construction Economics – An Introduction Stephen L. Gruneberg
Construction Law Michael F. James
Construction Marketing Richard Pettinger
Construction Planning, Programming and Control Brian Cooke and Peter Williams
Contract Planning Case Studies B. Cooke
Cost Estimation of Structures in Commercial Buildings Surinder Singh
Design–Build Explained D.E.L. Janssens
Economics and Construction Andrew J. Cooke
Environmental Management in Construction Alan Griffith
Environmental Science in Building, third edition R. McMullan
Estimating, Tendering and Bidding for Construction A.J. Smith
European Construction Procedures and techniques B. Cooke and G. Walker
Facilities Management, second edition Alan Park
Greener Buildings – Environmental impact of property Stuart Johnson

(continued overleaf)

LEEDS COLLEGE OF BUILDING
WITHDRAWN FROM STOCK

Housing Associations, second edition Helen Cope
Housing Management – Changing Practice Christine Davies (Editor)
Information and Technology Applications in Commercial Property Rosemary Feenan and Tim Dixon (Editors)
Introduction to Building Services, second edition E.F. Curd and C.A. Howard
Introduction to Valuation, third edition D. Richmond
Marketing and Property People Owen Bevan
Measurement of Building Services George P. Murray
Principles of Property Investment and Pricing, second edition W.D. Fraser
Project Management and Control David Day
Property Development Appraisal and Finance David Isaac
Property Finance David Isaac
Property Investment David Isaac
Property Management – A Customer Focused Approach Gordon Edington
Property Valuation Techniques David Isaac and Terry Steley
Public Works Engineering Ivor H. Seeley
Quality Assurance in Building Alan Griffith
Quantity Surveying Practice, second edition Ivor H. Seeley
Real Estate in Corporate Strategy Marion Weatherhead
Recreation Planning and Development Neil Ravenscroft
Resource Management for Construction M.R. Canter
Small Building Works Management Alan Griffith
Social Housing Management Martyn Pearl
Structural Detailing, second edition P. Newton
Sub-Contracting under the JCT Standard Forms of Building Contract Jennie Price
Urban Land Economics and Public Policy, fifth edition P.N. Balchin, G.H. Bull and J.L. Kieve
Urban Renewal – Theory and Practice Chris Couch
Value Management in Construction B. Norton and W. McElligott
1980 JCT Standard Form of Building Contract, third edition R.F. Fellows

Series Standing Order

If you would like to receive future titles in this series as they are published, you can make use of our standing order facility. To place a standing order please contact your bookseller or, in case of difficulty, write to us at the address below with your name and address and the name of the series. Please state with which title you wish to begin your standing order. (If you live outside the United Kingdom we may not have the rights for your area, in which case we will forward your order to the publisher concerned.)

Customer Services Department, Macmillan Distribution Ltd
Houndmills, Basingstoke, Hamipshire, RG21 6XS, England.

Construction Marketing

Strategies for Success

GE OF BUILDING L
ıst be returned on or befor
stamped below.

Richard Pettinger
University College London

© Richard Pettinger and Rebecca Frith 1998

All rights reserved. No reproduction, copy or transmission of this publication may be made without written permission.

No paragraph of this publication may be reproduced, copied or transmitted save with written permission or in accordance with the provisions of the Copyright, Designs and Patents Act 1988, or under the terms of any licence permitting limited copying issued by the Copyright Licensing Agency, 90 Tottenham Court Road, London W1P 9HE.

Any person who does any unauthorised act in relation to this publication may be liable to criminal prosecution and civil claims for damages.

The author has asserted his rights to be identified as the author of this work in accordance with the Copyright, Designs and Patents Act 1988.

First published 1998 by
MACMILLAN PRESS LTD
Houndmills, Basingstoke, Hampshire RG21 6XS
and London
Companies and representatives
throughout the world

ISBN 0–333–69278–0

A catalogue record for this book is available
from the British Library.

This book is printed on paper suitable for recycling and made from fully managed and sustained forest sources.

10 9 8 7 6 5 4 3 2 1
07 06 05 04 03 02 01 00 99 98

Printed in Malaysia

Contents

Preface

Other than quarrying for the materials that it uses, building and construction is the most primary of all industries. Even where people exist on standards of living very close to the borderline between life and death, some form of shelter and the means of moving around are essential if this pattern of life is to be sustained for any length of time. The industry itself is therefore both global and universal, truly reaching and touching every part of the world and all those people who live in it.

Throughout the world, there are great pressures on the industry. In the West, these concern especially the use of resources; effects on the environment; pressures for conservation, refurbishment and recycling as well as building anew. There are also special pressures – particularly cost – on those who commission publicly funded projects and activities. Enhancing the quality of life and the quality of working life, and all that this implies both for domestic and commercial building, and the infrastructure necessary, is a matter of fierce debate. Pressure groups and vested interests have learned to gain access to and command of the media and have succeeded in both raising the level of general concern and awareness, and in highlighting specific factors relating to particular issues.

Elsewhere in the world there are extensive drives for 'civilisation' – in whatever terms that is defined in the given location. In general, it refers to forms of urbanisation, commercialisation and pressures towards consumerism from those who have not hitherto enjoyed their full benefits (real and perceived) and the opportunity that these bring.

These pressures are compounded by 'the dash for work' on the part of the largest companies in the industry – the attempt to gain footholds in new markets wherever these can be found. The construction industry in Western Europe and North America was always traditionally domestic – it was almost unthinkable that foreign companies would be invited to bid, tender and contract for work, whether given the opportunity or invited expressly to do so – and certainly not at the behest of government policy or to contract for public projects. In the UK, this was compounded by extensive encroachment by European, Japanese and North American companies into the construction of major projects. More locally, the domestic house building and commercial construction sectors were hit by loss of consumer and client confidence in these markets. This has therefore resulted in turmoil – some have said a crash – in all parts of the industry. Companies have been forced to compete with foreign organisations, as well as domestic alternatives, on their own patch for the first time. This has occurred at the same time as there has been a radical downturn in public sector investment and the piloting of new initiatives such as the private finance initiative; and whether these will succeed in the future it is far too early to assess.

The companies involved have therefore both added to their own turmoil and also caused the ripples to be felt among smaller, regional and local operations as they have sought work elsewhere. 'Elsewhere' has meant both at home – where in some instances they have sought to force their way into those markets hitherto dominated by smaller domestic or regional operators; and abroad – where they have joined with varying degrees of trepidation, investment and commitment, the trek to the tiger economies of Asia and the emerging markets of the ex-USSR. Others again have sought to open up operations in the established markets of the USA and Australia; others, still, have sought niche opportunities in South Africa and South America.

Whether pursuing work in domestic markets or overseas, construction organisations have had to face the fact that their client groups now have a greater range of choice than ever before as the industry – like so many others – has become global and unprotected. In most, if not all cases, no contracting organisation bids for work unless it is determined to win it. The level of expertise in all those who tender for a given project is certain to be at least adequate. The client is therefore going to make its choice on grounds other than capability. The difference therefore lies in the ability to present the expertise in ways designed to persuade and influence the client to buy from one organisation in preference to others – and this depends to a great extent on the quality and professionalism of marketing.

Marketing is concerned with presenting products, services and expertise as client benefits. These are offered to best advantage so that an enduring and profitable relationship is engaged. Many parts of the industry do this extremely well. For example, building products are offered, both retail and wholesale, in ways familiar to the buyer and at known and accessible points of contact; specific items such as double glazing are offered through a combination of advertising, reputation and direct sales – also very successfully in many cases.

For the rest of the industry, the main problems are attitude and understanding. In previous times, there was always enough work around to ensure that a reasonable share could be undertaken by every company concerned and that this would support good levels of profits. Long-term, steady and stable relationships between contractors and clients were engaged – and this led to the creation of long-term effective relationships. Now, however, all this has changed.

Hitherto, there was always more or less a balance between the contractor and client interests, and this was founded on the knowledge and certainty that work would always be available. Now the industry is dominated by client groups. There are too many contractors chasing too little work – and these contractors come from all over the world as well as the domestic industry. The result is a downward pressure on price; an upward pressure on quality; an assurance of service; and an upward pressure on the ability to meet deadlines, deal with pressure groups and interests, undertake client liaison and manage any specific local factors such as traffic disruption or domestic and commercial inconvenience while projects are being undertaken. This requires a fundamental shift of understanding on the part of all those responsible for the direction and strategic purpose of construction companies.

Attitude is also all-important. In the past, companies of the industry concentrated on product and technological excellence supported by top quality staff. This remains important; but the difference is that these qualities have to be presented as client benefits given the great range of potential contractors from which clients are able to make their choice. It is necessary, therefore, for the attitude to be shifted from product excellence to service assurance – and this is only possible if those companies involved undertake major initiatives to ensure that they know their clients; know the pressures and drives under which they have to operate; and deliver the product, service excellence and expertise in ways that match client demands.

With this in mind, there is a single lesson that the construction industry needs to learn above all others from marketing in the consumer goods sectors. The most successful consumer goods companies in the world – Sony, Nissan, McDonald, Coca Cola – remain so because as well as ensuring the excellence and constant improvement of their products, **they market**. If they did not market, they would lose market share – and income, sales and profits – to those who continue to do so. The same applies to construction companies. Overseas companies that have made successful entries into UK markets have done so, not because their product or expertise is any better than the domestic industry (it is not); they have done so because they have been able to present their expertise in terms of client benefits and to engage sufficient confidence and trust to be awarded the work. **The difference therefore lies in their marketing capability.**

There are also lessons for the larger companies from smaller operators. It is still perfectly possible to find sole traders, small partnerships and small company jobbing builders who have high and continuing volumes of work. While they will not necessarily think in the precise terms used in this book (indeed, many certainly do not), it is absolutely certain that they will have built a local, enduring and current relationship within their particular area so that work keeps coming in to them. As well as producing work to high quality and standards, this means offering and delivering high levels of service and convenience, friendliness, confidence and reassurance – and all this means that their particular clients choose them above others.

The outcome of all of this is the need for the construction industry as a whole to bring itself up to the marketing standards of those who do it well; and to learn to apply the principles of successful commercial activity that are operated so effectively elsewhere, so that the best possible chance of securing work is achieved. And for those who say that the construction industry is unique – they are right, it is. So is every other industry. And, as for every other industry, marketing for building and construction means understanding and adopting the principles of marketing and applying them to your own particular circumstances. That is the purpose of this book.

Chapter 1 is an introduction to the principles and practices of marketing: what marketing is; what marketing is not; and forms the basis on which the marketing attitude and approach are adopted. Chapter 2 considers the customers' and clients' needs, wants and perceptions: the key elements of confidence and satisfaction; and more specific needs such as the ability to deliver on price (for public contracts) and quality (where price is not the main consideration). Chapter 3 is concerned with strategy – this is often mistaken for remote, corporate, head office activities; the reality is very different, although it is necessary to have a clear purpose in mind if this is to be supported with effective marketing activities.

Chapters 4, 5, 6 and 7 consider, in turn, the key marketing activities of product, price, place and promotion. Often referred to as the 4Ps or the marketing mix, it is the combination of each in particular circumstances and with particular customers and clients that provides the basis for choice. Each is therefore considered in some detail. Extensive reference is also made to the concepts of 'service marketing' – in which construction is seen as service delivery as well as product excellence; and 'relationship marketing' – because of the importance of the direct contact between contractor and client, the professionalism necessary, and the ability to deal with clients who are not construction experts as well as pressure groups, vested interests and other lobbies.

Chapter 8 is concerned with the need for information on the markets and client bases whom it is intended to serve. Chapter 9 then draws on examples where successful marketing is being carried out within the construction industry and the lessons that can be drawn for the rest of the sector for the future. It also looks in more detail at the need for attitude and understanding shift.

The book is aimed primarily at undergraduate and postgraduate students on building, construction, surveying, civil engineering and construction management courses. Many of these courses do now have marketing elements. Where these elements are present, this book is a key text. Where they are not, the book is nevertheless an essential reader in filling this particular gap.

The book will also be of value to those following professional membership routes – for example, of the Institution of Civil Engineers, the Chartered Institute of Building, Royal Institute of British Architects, Royal Town Planning Institute and the Royal Institution of Chartered Surveyors. It will be of great value to those learning the consultancy business – especially for those who may be required to work in overseas or unfamiliar markets at short notice.

It will also be of value to those already working in the industry, many of whom still do not understand why they continue to have to compete harder for an ever-diminishing volume of work or the range of steps that could be taken to ensure greater volumes of work: greater strike rates in terms of 'number of tenders per contract awarded'; and some of the pitfalls (as well as opportunities) associated with moving into new areas of activity and overseas markets.

University College London, 1997 RICHARD PETTINGER

Acknowledgements

This project was originally conceived because there was no core text available for the marketing elements of the BSc Construction Management course at University College London. Subsequently, it was extended further because there was also no core text available for the marketing elements of specialist postgraduate taught Masters programmes in construction management, civil engineering management and facilities and environment management.

I am therefore extremely grateful for the support given by Derek Beck, Patrick O'Sullivan, Bev Nutt and David Kincaid of the Bartlett, UCL in the pursuit of this project.

Many other people also helped. Ivor Seeley, Emeritus Professor of the Nottingham Trent University, provided an extensive review of the manuscript and made very many positive suggestions. Anthony Impey and Mark Genney were constant sources of inspiration and provided both insights and material. Ram Ahronov was also a constant source of support and guidance, and also helped with providing material. Stephen Gruneberg critically reviewed some of the ideas and examples.

I am very grateful to Malcolm Stewart at Macmillan Press; and to the undergraduate and postgraduate students of the Bartlett School of Architecture, Building and Town & Country Planning for all their support, advice and guidance.

1 Introduction

Marketing is the competitive process by which goods and services are offered for consumption at a profit. This definition should be seen in the broadest possible terms. It is about building a reputation and making sales for generating long-term positive and profitable business activities in specific markets. It means gaining credibility and reputation in order to establish effective and co-operative working relationships.

Quite apart from having sufficient expertise and products to offer for sale profitably, there is a requirement to generate images, reputation, confidence, positive attributes and expectations in order to ensure that these activities remain successful and productive.

The fundamental concept of marketing is that all business activities are in some way the relationship between oneself and one's clients. Marketing takes the view that the most important stakeholders in the organisation are the customers and that keeping them contented and satisfied, and returning for repeat business, is the first concern of all the staff. Marketing activities are thus central to the function of everyone in the organisation – and marketing (whether positive or negative) is carried out by everyone. Every interaction between the organisation, its staff, its clients and its environment has some input on the relationship between them.

In support of this, organisations devise marketing strategies and carry out ranges of activities designed to assess and satisfy customer needs and wants. They gather market intelligence, conduct market research and obtain customer responses. They organise product and service packaging, promotion, sales and distribution. They generate product and expertise awareness through advertising, image formation, differentiation, promotions, sales methods and techniques, and presentations.

Marketing is at the core of the strategic purpose of the organisation. Without a clear purpose of the nature of operations and the business that it is going to conduct and be in, it will have no impetus or clarity of direction. Marketing is an all-pervasive business concept. It is there to ensure that what is done is both effective and profitable. It is concerned with securing the continuity and long-term future of the organisation. It is the generation of and the bringing on stream of a steady supply of new and improved products, projects, initiatives and services that are suitable in both operational and quality terms for the markets in which the organisation operates or seeks to operate. It is about the development and identification of marketing opportunities that accord with the capability and mix of activities and expertise.

Marketing exists wherever there is choice. People have to be persuaded to use one facility, service, product, project or item in preference to others. Even where on the face of it choice is very limited – there is at present no real choice concerning the consumption of electricity or water – people still have choice in terms of whether or not to switch the light on or how often to have a bath, do the washing or drink water – and in drinking water, whether from the tap or from the bottle.

Successful and profitable organisations understand this. They know that it is not enough simply to have an impressive track record or to rely on general perceptions that 'everybody knows who we are and how good we are'. Indeed, the most successful organisations, those that are well known for their track records and product quality throughout the world, continually reinforce and re-present their products and services through continued attention to marketing and its media.

Some construction companies are expert marketers. More generally however, it is clear that there are lessons to be learned by companies in all parts of the industry from other sectors. For example, there is no reason to buy Coca Cola – as a soft drink it is expensive relative to other offerings (above all, other colas) and in times where there is supposed to be a heightened awareness of what is good and healthy to eat and drink, a 50 cl tin contains 550 calories. The main component of Coca Cola is sugar. And yet people do buy it in their millions. On average one person in eight on the planet drinks a Coca Cola product everyday (the company has sales of one billion per day and there are eight billion people in the world). The company has the largest volume of product sales in the world. The word Coca Cola is the second most recognised in the world (after OK).

The difference lies in marketing. Coca Cola has identity, brand strength, confidence, expectations, images and associations which it has generated over a long period of time – and most important of all, which it continues to generate now and for the future. All this is supported by:

- Global and general advertising, promotion, hook-up and association campaigns and initiatives that ensure strong positive brand strength overall.
- Localised and specialised promotions, hook-ups and association campaigns and initiatives that reinforce the positive images in response to specific economic, social, communal and ethical considerations.
- Pricing levels that are set for the locality and outlet in question.
- Distribution and presentation networks – shops, bars, restaurants, stalls, kiosks – supported by manufacturing and delivery systems that ensure the product is readily and continuously available.
- Continually presenting the advantages and benefits of using the particular product.

There are fundamental differences between marketing consumer goods such as Coca Cola and marketing construction and its activities. Yet there are clear similarities and the overall purpose is the same. If a construction organisation wants a distinctive presence, it has to invest in the marketing activities that will secure this. In any case, it must market itself in the localities in which it exists and works – and it must continue to do this in terms acceptable to these. It must set prices and charges that are acceptable and achievable. And it must also have distribution networks to support the promises that it makes or infers when bidding or tendering for work; and this includes access to subcontractors' plant and equipment, materials and resources.

There are major differences between the markets served by the construction industry and consumer products. These include few buyers – compared with consumer goods sectors; and infrequent purchases – compared with consumer goods sectors. The following should also be considered.

THE SIZE AND NATURE OF BUYERS

Most of these are institutions of one sort or another, including companies and organisations who commission project work, new developments, refurbishment and upgrades; and consultancy and problem solving.

These also include governments and their departments, and other international institutions such as the United Nations and NATO. Work may also be commissioned by institutions such as the International Monetary Fund and other global assistance activities.

The same thing also applies at more localised levels and this includes the commissioning of hospitals, schools, social services institutions, roads and infrastructure projects by regional and other public bodies. Housing developments are commissioned as the result of public and local planning inquiries. Public service bodies and private sector corporations commission refurbishment and premises improvement work. Private individuals also commission this sort of work – and again, the basis on which they do this is that 'it will only have to be done once, for the foreseeable future at least'.

NATURE OF DEMAND

Demand for construction projects may be:

- Derived, which means that work is created in response to political, economic, social and community demand for facilities such as shopping malls, schools, colleges and hospitals.
- Creative, whereby these facilities are commissioned as part of new development or regeneration programmes to try and encourage further economic activity.

- Progressive, which concerns refurbishment and upgrading; and this also applies to the purchase of building services and building materials at a consumer level.
- Innovative, where it is proposed to transform a whole location from one social form (e.g. rural) to another (e.g. urban).

PRIMACY OF DEMAND

It is often said that the construction industry is now the first to suffer at the start of a recession and the last to benefit at the end when the particular economy starts to grow again. This varies between countries and regions and some governments and government departments take the view that the industry can be provided with work to generate local wealth which, in turn, will enable those who work on the projects to spend money on other goods and services and increase tax revenues. Others take the view that the construction activity will follow in the wake of consumer and service led economic confidence.

From this the following scenarios are extrapolated:

- Where other economic activity is to be generated on the back of publicly commissioned construction projects, the emphasis of the industry is in generating and producing pioneering work, new designs, competitions, technological advances and supported by extensive capital investment.
- Where activities are to be led by increases in other economic activities, the emphasis in the first place is in the upgrading of existing facilities, refurbishment and rejuvenation of existing environments; and the capital intensive pioneering work will follow on subsequently.

CUSTOMER–CLIENT RELATIONSHIPS

In the supply of consumer goods, this consideration is distant, peripheral and transient. The relationship between consumers and suppliers is founded on product, branding and customer loyalty. This is generated through image, identity and confidence based overwhelmingly on advertising, promotion and differentiation. The result, at its best, is a strong customer–supplier relationship based around the product itself which is, nevertheless, remote from the supplying organisation (e.g. Coca Cola customers have great loyalty to the product, little or no identity with the company – and very little knowledge of the company's activities, institutions, ways of doing business, ethics and so on).

In construction, this form of relationship exists only at the peripheral retail end of the market where building products, house fixtures and fittings and some plant equipment are sold over the counter.

Overwhelmingly in the industry, effective contractor–client relationships are based on continued extensive and influential, personal and professional relationships between supplying and demanding organisations. Contractors are invariably expected to produce specific and distinctive products, projects, product mixes and project proposals – and this can only be truly achieved where there exists a high level of mutual knowledge – and high levels of confidence that arise as the result of this knowledge and understanding.

Clients therefore commission work on the basis of professional knowledge and understanding rather than perceived product quality – and this is a fundamental difference with consumer sectors because it affects the entire emphasis of effective construction marketing activity. Many organisations – especially architects, designers, planners and specialist builders – expect or perceive that high quality of work, supported by an impressive track record, wide field of historic, recent and current performance, is enough. Often, and increasingly, this is no longer the case – clients commission work because expertise is supported by a continued high level of attention to all their specific needs.

Moreover, in many cases the work is commissioned by people – institutional boards of directors, governors of public institutions, politicians – who have no particular knowledge or understanding of the construction industry. These people will have taken extensive advice from those who do know the industry; and once the work is in progress, continuous relationships covering project liaison, the purchase of equipment and primary materials, the commissioning of subprojects and subcontracts that now become necessary are all carried out between specialists, professionals and experts.

ABILITY TO CHOOSE

Clients and client bases also have an ever expanding field from which to make their final choice of contractor.

At the macro end, the largest construction and civil engineering organisations are now truly global and may, and often do, work anywhere in the world. Many major projects now have to be joint ventures simply because no single organisation has the capital, human resources, technological expertise or cultural know-how to do the job on its own.

Raw and primary materials for the industry are now available from many parts of the world. Control of supplies is no longer as easy as it once was, and this applies especially to Western European and North American companies that have hitherto been able to guarantee supplies from their own mining, quarrying and distribution subsidiaries.

Expertise is also evermore widely available. This has arisen as the result of many influences: there are now both company based and independent project management, geotechnical design and planning divisions and agencies; public service planning and related functions that have either been reconstituted or else completely privatised, are now able to offer their expertise on the open market; cultural barriers that historically led to the domestic industry being used for domestic and, above all, publicly funded projects have changed.

The wider and increasing availability of expertise has come at a time therefore when clients and commissioners are taking a much broader and uninhibited view of where they can get work done, and with this comes the capability to ask, during tendering processes, for a much broader view of the finished project (see Summary Box 1.1).

SUMMARY BOX 1.1 Economist Advertisements: The Project Seeking Format

'International Construction Consultants (ICC) is a leading German construction, civil engineering and project and facilities management consultancy. ICC identifies, executes and evaluates projects worldwide for and on behalf of EU and non-EU government, international funding agencies and private sector clients.

For upcoming projects we seek experts in: engineering; civil engineering; business processes; quality control; plant management; geotechnics; quantities and quantity surveying; finance and accountancy; marketing; economics; project management; rural and urban development; institutional strengthening.

All applicants must have at least three years of international working experience; professional qualifications and experience in any of the stated fields; fluency in English (the normal working language of ICC projects is English); other European and Asian languages have a distinct advantage.'

'OPM is an expanding consultancy firm which supports public sector and international clients with high quality, economic, organisational and specialist construction industry services. It seeks staff and associates with the following: policy experience in public service contracts; public service reforms; public finance; rural and urban regeneration finance; EU development programmes; the provision of health, education and social facilities.

Specialist/technological experience in the following fields: project management; surveying; contract management; cross-cultural and transnational activities; building and engineering quality assurance; water, drainage and other main subcontracting works.

All applicants should have sound academic training with good higher degrees and professional, consulting, analytical and advisory experience.'

This form of approach to construction projects is based entirely on building a successful network of experts that are to be called on at the points required during activities. The expertise of the networker lies in drawing specific organisations, for specific parts of the work that are both capable and willing to work in this way. It is an increasingly familiar format in the prequalification and marketing of construction work capability in South East Asia, the former USSR, the former Yugoslavia and South America.

CLIENT CONCENTRATION

Historically, client concentration in the construction industry was always seen as being geographically based – especially at the macro end; and otherwise regional or local. This worked well where there was sufficient (or even too much) work to be shared out on this basis. Now however, an over-supply of work in one region, nation or locality attracts players from elsewhere in the world where demand is slack and this has led to the view that it is possible to see client concentration in a variety of other ways including:

- **Expertise**: this approach concentrates on parcelling up clients according to the nature of work that they commission, e.g. water supply (Middle East, central ex-USSR, Brazil, South Africa, Argentina); transport infrastructure (Western European Railways, the former USSR, Australia); urbanisation (China, Indonesia, Malaysia, South America, South Africa).
- **Process**: this approach is more long term (even by the special constraints of the industry); and it concentrates on extending the precontract period and discussions to get the client to acknowledge the true extent of its requirements (see Summary Box 1.2).

SUMMARY BOX 1.2 **Construction as Process**

The Indonesian government requested bids to build a cement factory near Jakarta. An American firm made a proposal that included choosing the site, designing the cement factory, hiring the construction crews, assembling the materials and equipment, and turning over the finished factory to the Indonesian government.

A Japanese firm, in outlining its proposal, included all of these services plus hiring and training the workers to run the factory, exporting the cement through their trading companies, using the cement to build needed roads out of Jakarta and also using it to build new office buildings in Jakarta.

Although the Japanese proposal involved more money, its appeal was greater and they won the contract. The Japanese viewed the problem not just as one of building a cement factory (the narrow view), but as one of contributing to Indonesia's economic development, i.e. a form of added value. They saw themselves not as an engineering project firm, but as an economic development agency. They took the broadest view of the customer's need. This is true system selling.

(Kotler, 1993)

- **Service**: this approach concentrates on applying some basic principles of consumer marketing. Expertise, capability and willingness are presented in terms of flexibility, dynamism and responsiveness demonstrably underpinned by a capital, financial and human resource base that can operate anywhere required.

- **Corporate citizenship**: this approach may also be called relocalisation. It concentrates on relationships with the locality/community in which the organisation is, or is to be, located. It addresses reasons other than the fact of its location as to why it should continue to be used for work. This approach means in effect becoming a community facility – for example, offering work experience placements to schools and colleges; sponsorship and support for local events; use of companies' facilities for meetings (e.g. Women's Institute, Scouts, Guides, Brownies and Cubs). All of this is supported by good-quality, targeted public relations and extensive local media coverage (see Summary Box 1.3).

SUMMARY BOX 1.3 **Channel Tunnel**

In 1986 the contract to build the Channel Tunnel was awarded to Eurotunnel. During the inception period, Eurotunnel conducted extensive local public relations. The company concentrated above all on the future – producing high-quality educational materials for use by children in schools and colleges of all ages. The first facility that the company actually built was an exhibition centre – detailing the project from its earliest conception (in 1806 during the Napoleonic Wars), through other failed projects to the present initiative. It designed and developed a resource, archive and library facility. It opened its doors to local schools, social services and community groups. It trained its public relations and other point of contact staff to be positive and dynamic in its dealings with anyone from the local community (or anywhere else for that matter) and to make sure that any queries, questions or concerns that they had were answered. The company also went to a lot of trouble to get high-profile, high-quality and regular local media coverage.

The result was that the company very quickly had an extensive, positive local reputation – whatever the community thought about the actual construction of the Tunnel (and at the time there was substantial opposition). Eurotunnel's emphasis was to concentrate on the benefits that the facility would bring to the community and because it was positive and dynamic about these itself, initial understanding and confidence were generated and a much broader acquiescence (if not outright acceptance) followed in its wake.

- **Cost leadership**: for clients (overwhelmingly those from central and local government institutions) that concentrate on price, the ability of contractors to present and demonstrate their cost leadership (and therefore the ability to produce effectively at the lowest possible price) is the key.

  Cost leadership also means taking the other factors indicated above into account – it is not an end in itself. However, it does indicate the need for the distinctive point of view of the capacity for long-term price advantage to be presented to prospective clients in this context. For major projects, it is also necessary to take into account the enduring, potential and likely costs of the finished facility over the duration of its useful life.

- **Repeat business**: there is no longer any guarantee of repeat business – either from a particular client or within a particular sector (e.g. public service; design and build) simply because contracting companies are casting about evermore widely for work. Part of construction marketing activity has therefore to concentrate on carrying out an initial piece of work as effectively as possible on the grounds that, if this is not done, there is no or little chance of repeat business being generated.
- **After-sales**: quite apart from anything else, clients can demand after-sales support and maintenance activities as an integral part of the prequalification and tendering process. Contracting organisations have therefore to develop this aspect of their activities – and this must include both capability and willingness. Moreover, under private finance initiative and similar arrangements whereby the contractor recoups its costs by charging for the use of the finished facility, it has become necessary to acquire operational and facility management expertise as well as construction.
- **Length of contract**: the general rule is that clients now expect contracts to be completed in the agreed timescale. Historically, this has not always been the case – poor industrial relations, bad weather, unavailability of supplies, delays by subcontractors and late supply of details by architects have all been blamed at various stages for the inability of companies to complete work on schedule. From the client's point of view, none of these are acceptable any longer – the requirement is placed on contracting organisations to sort out their labour relations problems, sort out their problems of supply and take steps to minimise the effects of bad weather themselves.

Table 1.1 compares consumer and construction marketing.The purpose of this summary is to indicate the groundwork necessary for effective marketing activity. The key clearly lies in the nature of long-term personal and professional contact that is required in the marketing of construction activities; and this is distinct from the most effective form of contact for consumer goods, which is mass advertising and promotional activities.

There are some other useful contrasts:

1. **Responsibility and consequences**: the consequence of loss of sale of a tin of Coca Cola is minimal/negligible; and this would only be affected if product quality was known or perceived to have been reduced.

Table 1.1 *Consumer and construction markets compared.*

Consumer goods	*Construction*
• Many buyers	• Few buyers
• Cash/credit	• Capital
• Small purchases	• Large purchases
• Individual buyers	• Corporate/public buyers
• Low consequence	• High consequence
• Easily disposable	• Not easily disposable
• Low responsibility	• High responsibility
• Universally available	• Specialist availability
• Short-term benefits	• Long-term benefits
• Low obsolescence	• High obsolescence
• Non-reusable	• Reusable/refurbishable
• Low technological consequence	• High technological consequence
• Low direct contact	• High direct contact
• Easily replaceable	• Difficult/impossible to replace
• Frequent purchases	• Infrequent purchases
• Instant purchase	• Long lead time to purchase
• Low contract value	• High contract value
• Non-specialist purchase	• Specialist expert purchase
• Individual benefits and satisfaction	• Considered purchase
• Direct benefits and satisfaction	• Indirect benefits and satisfaction
• Satisfaction comes from immediate usage	• Satisfaction comes from subsequent usage
• Short useful life	• Extensive useful life
• No/low investment required	• High investment requirement
• Loose supplier–customer relationship	• Close contractor–client relationship
• Transient supplier–customer relationship	• Extensive, professional contractor–client relationship

Loss of a single sale to a construction company (from any part of the industry) has a direct financial bearing.

2. **Repeat business**: the keys here are the frequency with which repeat business is generated. For the construction industry 'repeat business' and 'long-term client relationship' may involve no more than two or three contracts per decade.
3. **The buying chain**: consumer goods are decided on, bought and used by individuals and their immediate circle. For construction, the decision, contracting, production and usage are often all carried out by different groups.
4. **Doing a good job**: those who produce consumer goods receive negative feedback in terms of increases/decreases in sales, customer complaints, use of competitors in preference and so on. Construction professionals in all disciplines may, and often do, carry out their work to the

best of their expertise and belief, and to the satisfaction of the client, only to find that the end-users may neither like nor want the facility.

MARKETING OBJECTIVES OF CONTRACTORS

These are:

- To tender and bid for work in ways that relate their capacity, expertise and willingness to the client's needs and wants.
- To generate a sufficient volume of work to support the size, scope and scale of the organisation.
- To generate the next job or round of work; and to generate an adequate return for their work.
- To compete with other players in the sector, discipline or expertise on the basis of particular strengths.

The following have also to be addressed:

- Promoting and generating confidence and expectations (see Summary Box 1.4).
- Building and reinforcing the organisation as a known sectoral player and sectoral brand.
- Generating customer loyalty and repeat business.
- Identifying new areas where the distinctive expertise may be applied successfully and effectively, and carrying out the research and groundwork involved in making sure that (a) those areas do offer genuine opportunities, and (b) that the new potential client base is aware of the new organisation in the field.
- Preaching to the familiar and converted through general industrial, financial and trade media.
- Preaching to the unfamiliar and unconverted through industrial advertising, product placement, use of the specialist media of the client base.
- Servicing the client base through substantial, expert and influential sales and client liaison teams; and generating substantial familiarity in new potential client bases.
- Generating a positive and confident public relations, media and community liaison facility.
- Using specific documents and outlets – above all, the annual report, recruitment advertising, site presentation – to best advantage as these are the initial points of contact and familiarity to the public at large, as well as more interested parties and stakeholders (see Summary Box 1.5).

SUMMARY BOX 1.4 **Luxury and Exclusivity**

'Being the best' has still to be marketed. For example, Rolls Royce cars are extensively marketed and in very distinctive ways – through media coverage on specialist motoring television programmes and in magazines; news coverage of royalty and show-biz stars who are seen driving them (and being driven in them). The car has a very distinctive logo – the figure on the bonnet. Product placement – using the cars in films and glamorous television shows – is also widely used. And this is reinforced by the very high price and by the performance of the cars in relation to customer needs.

This applies to expensive and exclusive design, architecture, planning and technological consulting services. People have to know who is the best, the most exclusive – and why. This is achieved through the same means and methods as those used by Rolls Royce – extensive specialist media coverage; reference by national and TV news media when covering their expertise (and especially when controversies occur); and product placement wherever possible (e.g. using particular buildings or locations, either for drama or documentary series, or business programmes). The most prestigious organisations are also those that are invited to tender for public competitions. Entering into these competitions is part of the marketing process – and whether or not the particular design is the winner, it can still be exhibited and given the media and public relations coverage.

Other marketing objectives

From the contractor's point of view, the following factors have also to be considered.

Repeat business

The object here is to create:

(a) customer loyalty, whereby previous customers return to you because of the good job that you did for them in the past and where what you now offer continues to command the same perceived standard of quality, value, workmanship and expertise;
(b) brand loyalty which is an association of you with all that is best and required in your particular line of work.

The two are different sides of the same coin. For example, when they need (want) a cold soft drink, more people buy Coca Cola than anything else. This runs along twin tracks as follows:

- 'I am thirsty, therefore I'll have a Coke' and

SUMMARY BOX 1.5 **The Annual Report**

The Annual Report produced by public limited companies and others is the major source of published information generally available. In some cases it is the only source of published information available. Annual Reports are available for scrutiny by the public at large and copies have to be made available upon request, whether the requester has any direct or indirect involvement in the company or not. It therefore has at the very least a general awareness – and therefore marketing – function.

More specifically, Annual Reports are scrutinised in depth by:

- shareholders and their representatives, who want to know how well the company is doing, how safe their investment is and what their dividend is to be;
- staff, who want to see the company's own published analysis of the current state of performance and prognosis for the future;
- financial markets, journalists and analysts, who pronounce in finance sector specific terms on short-term future and confidence;
- industry sector specialists, who pronounce in industry sector specific terms on short, medium (and sometimes also long)-term future and confidence;
- students of the particular sector and of the company in question – and this includes professional bodies and associations, trade unions, research organisations, universities, colleges and schools;
- existing and potential customers and clients who scrutinise the report in search of evidence of what is important to them – existence of financial resources; long-term capability; projects undertaken and completed; work in progress; where the company's priorities lie; and an indication of the quality of service that they might expect.

In addition to this, as a marketing tool, the Annual Report operates at further levels as follows:

- it reinforces (or not) everything that the clients and potential clients have been told by the company in the recent past;
- it indicates general levels of confidence and expectation; and this is reinforced in turn, by stockmarket activity;
- it provides a summary of the company's actual (or near actual) financial position which again reinforces (or not) client and potential client understanding;
- it enables the organisation to present its strengths, advantages and achievements in their best possible light;
- it enables professional, academic and research bodies to add to their fund of knowledge about the company and about its sector.

- 'I'll have a Coke, therefore I will no longer be thirsty'.

So function (the former) and association (the latter) are both present and both have to be addressed. The outcome of this in construction terms is as follows:

- 'We need a factory designing, therefore we will use architecture practice X' and
- 'We will use architecture practice X, therefore we will have a good factory'.

While the marketing approaches are necessarily very different, the output is the same.

Broadening the client base

This means finding out who is commissioning work that you are both able and willing to do; meeting them; gaining their confidence; matching your capability with their requirements; and presenting it so that they want what you have to offer and will choose you ahead of the competition. This means above all, gaining a sufficient understanding of the potential client's needs and wants so that your capability can be presented as client benefits. This needs thorough and meticulous groundwork and is followed by individually targeted marketing (see Summary Box 1.6).

SUMMARY BOX 1.6 **'Why Use Us?'**

In simple terms, a substantial part of the process of developing an effective approach to marketing is in answering the following questions:

- Why do people use us?
- Why do people continue to use us? and
- Why do people not use us?

The answer to this is based on knowing and assessing the following:

- What do we do better than anyone else?
- What do we do better than most?
- What do we do worse than most?
- What do we do worse than anyone else?

and pinning these down as exactly as possible. The purpose is to get beyond self-perception and bland generalisation, acceptance and general awareness. Weaknesses are acknowledged, neutralised or remedied. Strengths become the starting point for effective marketing.

This is then to be seen in the context that what an organisation actually does best may not have penetrated the perceptions, knowledge and understanding of the client base. What it does best may not be important to the client base. At least understanding this gives pointers to the need and capacity for future market, marketing and organisation development. From this, an accurate and informed assessment of where and why the competition can be beaten – again, in terms required and understood by the client base – can then be understood.

Information

No contact with the market should ever be wasted. However negative a response from a potential client or ex-client may be, it always adds to the fund of information available to the organisation and this should be assessed, evaluated, judged and stored on its merits. The greater the fund of information, the greater the effectiveness and targeting of market activities.

Relationships

Organisations build relationships and networks with each other; and individuals acting in the name of the organisation do the same. Personal and professional relationships therefore exist, as well as the impersonal relationship between the contracting and client organisation. Every contact at every meeting is an opportunity for enhancing this, developing confidence, emphasising the positive, showing commitment and placing the client or potential client interests at the centre of things (see Summary Box 1.7).

Off duty

Preaching perfection, nobody should ever be completely off duty. Professional gatherings, courses and seminars all clearly offer opportunities for networking. So also does membership of sports, leisure and social clubs, and some organisations are sufficiently aware of the potential of this to pay the subscriptions of staff members who join them. And this may, and does, also extend to domestic dinner parties!

In practice, organisations are increasingly taking the view that nobody is ever completely off duty. There is a prevailing collective mood in the UK that anyone who does anything that may possibly bring their organisation into disrepute through their private actions and behaviour is considered a potential candidate for dismissal or discipline. As examples – a member of staff at a privatised electricity company was dismissed for embezzling the funds of the Brownie pack of which she was a custodian; a building society dismissed an employee for being involved in a nightclub brawl; a multinational car manufacturer introduced a smoking ban in all its company cars, and disciplined a middle manager for smoking in his car while he was on holiday.

All these took place in south-east England in 1996; and while none involved the construction industry, the message is clear: if the negative applies, so does the positive – which is that reputation, contacts and orders can be gained in any situation.

SUMMARY BOX 1.7 **Relationship Marketing**

In the past few years what marketing is about has been redefined again. Older ideas focused on techniques for winning new customers. The new approach says that the best route to growth and profit is to concentrate on keeping existing customers happy. This concept is called relationship marketing. It is having a dramatic effect on how companies do business.

Relationship marketing was kicked into gear by studies from consultants who calculated the remarkable value of loyal customers. They showed that loyal customers were assets with enormous lifetime values. Loyal customers are more profitable because they buy more of the companies products, require less time recommended to others and are less sensitive to price. In contrast, the cost of research, advertising and promotion is a very expensive way to grow. The average company loses 10% of its customers annually. Consultants, Bain & Co., found that if companies could increase customer attention by 5% they could massively boost profits.

These findings emphasise the importance of continually measuring customer satisfaction and following up on why customers defect. Instead of assessing managers on profits, companies would do better measuring them on customer satisfaction and loyalty. Current profits result from past decisions; tomorrow's profits depend on satisfying today's customers. It also makes sense to encourage customer and client openness – even encouraging them to complain. This way companies get a customer or client view of problems and, by addressing these, can keep their business. The key to creating customer loyalty is in the selection, training and motivation of front-line staff – and for the construction industry, this means attention to those who are going to work extensively with customers and clients on the gaining and maintenance of work. It also means paying attention to the quality of initial response – and this includes telephone contacts, public relations material presentation and speed and convenience of access.

Source: Peter Doyle – 'From the Top' (*The Guardian*, 18 January 1997)

Indeed, at the jobbing end of the market, relationships and work are generated through conversations over a drink, chance remarks, conversations in the street, gossip between friends and neighbours.

In summary, the overall objective is that marketing becomes an individual, group, organisational and corporate attitude. It is a state of mind that recognises, seizes and maximises any opportunity that may lead directly to business, that generally familiarises, or that may do no more than plant the seed in the mind of the recipient. And while not all jobbing builders, double glazing and building products sales people may actually think in this way, it is undoubtedly what they do if they are successful. And the very best know when to talk shop and when to back off and talk about something else!

2 Customers, Consumers and Clients

Marketing in construction is concerned with presenting each of the following as client benefits and enhancing advantages of doing business with the contracting organisation in preference to all the others that are available and willing:

- The capability, expertise, technology and resources that it has available; the capacity of these to support the length, scale and scope of activities required.
- The capacity to carry out work and produce the finished product in the client's best interests and to meet the client's specific requirements.
- The capability to produce work in such a way that the reputation of the client is not adversely affected; and that the client's reputation is enhanced as the result of taking delivery of the finished product, facility or specialist service.
- The capability and willingness to service the contract in order to build a positive contractor–client relationship as required.
- The capability to envisage the project and present the contractor's view of the client's needs in ways that meet and match the client's own projections.
- The ability to deliver the finished project or service to deadline, cost and quality.

CUSTOMERS, CONSUMERS AND CLIENTS

It is useful to distinguish the following:

- **The client:** commissions the project or activities and enters into the contractual relationship with the contractor; and entering this relationship is often supported by the advice of professional consultants, both independent and in-house, and also expert internal functions.
- **The customers:** pay for the project upon completion and sometimes into the future (for example, if further work is found to need doing or is in response to unforeseen and uncontracted circumstances, or as the result of an extended service agreement).
- **The consumers:** use the project now and possibly for its useful life; it should also be noted that consumer groups change – for example,

following refurbishment and upgrading activities long after the initial project has been completed.

Customers, consumers and clients may be the same or they may be a combination (the client may be the customer; customers may also be users; and so on). In many cases, it is a fine distinction. It is important at least to note it, as although each is very different, everyone's needs have to be satisfied.

CLIENT REQUIREMENTS

These are as follows, though the mix and emphasis clearly varies between size, range and volume of work and between contracts:

- Ability to deliver to the agreed price, timescale and quality.
- Adequate liaison during project inception, construction and completion.
- Early warning anticipation systems for genuine problems and unforeseen circumstances.
- Presentation of construction industry expertise (from whichever discipline) in terms of benefits and satisfaction to clients.
- Professional support during critical phases, e.g. public inquiries and dealing with pressure groups and lobbies; and for managing public contracts, managing and addressing the concerns of politicians and project governors.
- Limited liaison with anticipated consumers, users and community.

Different clients may put specific emphases, especially on the following:

- Public contracts are driven overwhelmingly by price. In the short term anyone really wishing to do this work has therefore to be prepared to compete on price alone if necessary, and to fit everything else in around this.
- Private sector contracts are driven overwhelmingly by quality, value and deadline for which price will be less important until it is agreed – but once agreed, it is normally immovable. Private sector clients expect the same attention to quality, value and deadline, supported by the necessary level of service, which they give in turn to their customers.
- Some clients have their own view of the balance of design, function, materials and durability. Others leave this – especially at the outset – entirely in the hands of the architects and contractors. This is either because they themselves are not quite sure what they do want or, because while they do have a good idea, they want to see if it can be improved upon.

The onus is therefore on the potential contractor to find out the client's actual requirements in advance of any tender or firm bid for work. The only way to do this is to engage in discussions with those at the client organisation who have real influence in the commissioning of work. Conversations between contractor and client public relations departments are much more comfortable; they are also lightweight and insubstantial – and any work that does come the contractor's way will be in spite of this form of contact and not because of it.

CLIENT EXPECTATIONS

Client expectations are based on the following:

- **Price, quality, value, volume, time**: the contractor will produce and deliver the promised facility according to the price, quality, value, volume and time mix agreed and contracted at the outset.
- **Suitability**: that which was proposed at the outset is actually what is delivered at the conclusion. Often this is based on the most general of perceptions only, though it can be addressed in part at least by computer drawings and projections, and also small-scale models and mock-ups. It is also addressed in more detail during client liaison activities while the work is being carried out.
- **Responsibility**: clients do not expect to hear excuses for late work, however 'valid' these may be.
- **Problem solving**: at the outset of a contract, contractors and clients establish, as far as possible, who is to be responsible for specific problems as they arise. This is essential. It is extremely bad marketing (and therefore bad business) to get involved in claims and counter-claims. This is exacerbated where potential problem areas are not clearly identified at the outset.
- **Confidence**: confidence is based on integrity, credibility and expertise; and on demonstrating that what is promised and implied is delivered in fact.

A critical step on the path to successful marketing lies in the ability to know and understand what the client's expectations are, where their priorities lie (e.g. in UK public contracts at the end of the 20th Century, the first priority is the price). The best way – some would say the only way – to achieve this successfully is to go out and meet with potential clients and ask them.

It is necessary to recognise that client expectations change, especially over long and expensive contracts. This may be outside their control, e.g. public sector clients may have their budgets cut unilaterally by their political masters. It may also be well within their control, e.g. 'this is not quite what we had in mind', and it is for this reason that effective and continuous client liaison is so critical to project success and forms a critical part of a project as well as marketing management.

Clients expect that everyone involved on the contractor side is there to respond to their every need. It is therefore essential that everyone who has contact with the client has full and complete understanding of the project purpose, style and methods of work and the intended outcome. Bad client liaison and management are the construction marketing equivalent of a bad TV advertisement to the fast moving consumer goods sector – and the result is loss of confidence, the desire not to be associated with a particular item and the beginning of the search for alternatives. Fast moving consumer goods can be disposed of easily – and so can companies in the construction industry. It just takes longer. Once confidence is lost or damaged on the client's side, it takes extensive and lengthy direct marketing activity to get it back.

Finally, clients expect their projects to be loved for all time. Where this is not the case – where the project is badly received by the community or final user group – a part of the negative response always gets back to the contractor.

TENDERS AND BIDS

In marketing terms, tenders and bids are submitted with one or more of the following purposes:

- To obtain the work on the basis of capability, expertise and quality of finished product.
- To obtain the work at any price.
- To obtain the work on the contractor's own terms, e.g. they will do it provided that everything is suitable to them.
- To obtain the work if the known preferred contractor cannot do the work (i.e. to become 'first reserve').
- To obtain the work knowing that if a good job is done, it will lead to the real prospect of a long-term relationship with the client.

Tenders and bids may also be submitted in general terms:

- To demonstrate capability to the given field and to the given client base.
- To expand the potential client base and field of work.
- As part of the general marketing, public relations and familiarisation process.
- In response to both general and specific requests to tender.

In marketing terms also, the tendering process starts with the following questions:

- From which of the above points of view are we tendering?
- Why are we seeking to gain this work?
- What are the consequences of getting it?
- What are the consequences of not getting it?
- Are we demonstrating our expertise to best advantage and in terms acceptable to the client?

This is the support for addressing the operational aspects of the tender, whether or not there is a legalised, standardised or expected format. There may also be contract compliance elements. Whatever the constraints, the final pool of tenders is certain to consist of those who are capable and willing to do the work, and to work within its given guidelines. The difference is therefore in the quality of marketing.

Companies may and do bid for work that they know they have little or no chance of getting. There are two main reasons for this. Either the client asks them to bid as a comparison so that they can see any difference of approach or possible alternatives. Or the contractor tenders anyway, using the process as part of its total marketing, public relations, awareness and image raising efforts. In either case, these questions have still to be addressed in case the client changes his/her mind and awards the contract to the particular company. Nothing is more frustrating, aggravating – and therefore downright bad marketing – than to be awarded a contract only to turn it down on the grounds of present lack of capacity.

From this, a picture of the marketing of construction may be built up (see Figure 1.1 on page 22).

This is a summary of the relationships, expectations and pressures that have to be accommodated in effective marketing of construction industry activities. The main emphasis is on relationships – and the consequent necessity to build relationships through direct personal and professional contact.

It should also be noted that there is a difference between capability and willingness, and the ability to satisfy everyone involved on both of these grounds; for example, the client's representative may be very willing to use a particular contractor based on knowledge of the contractor's capability – but this may not accord with other client organisational pressures (e.g. boards of directors, boards of governors) who may be unwilling to use the particular contractor for other reasons.

CLIENT RELATIONS

Good client relations are essential for the success of any business or activities. Good customer relations are based on finding out from customers and clients what their needs are, before and after-sales; supporting

LEEDS COLLEGE OF BUILDING

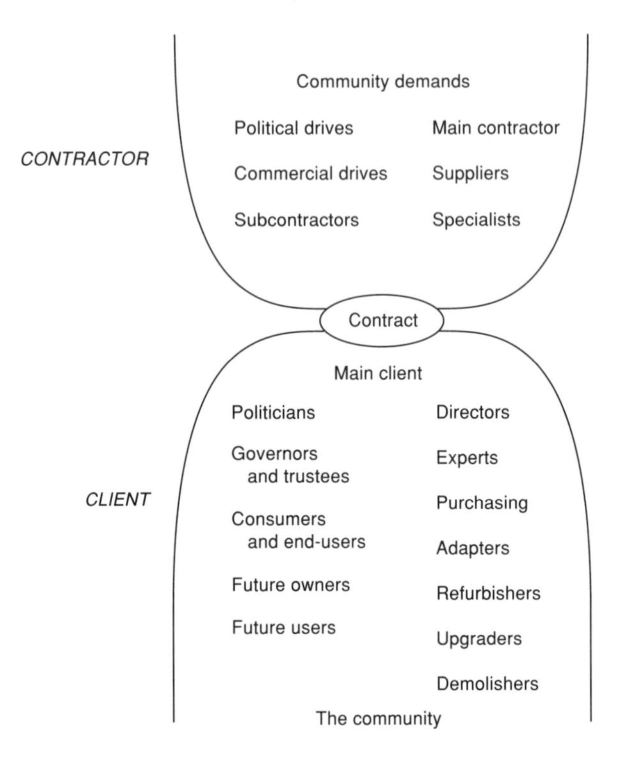

Figure 1.1 Marketing construction: the hour glass.

activities with high levels of service; developing new products and methods of service that accord with customer and client demands; and keeping in regular contact to ensure the continued ability to meet demand.

Bad client relations are normally due to poor service; inability to deliver that which was promised or contracted; a lack of interest in the future needs of the client; insensitive pricing, credit control and other cost-related elements; lack of personal and professional contact; unprofessionalism of contacts.

A balance has to be struck between each. This is because over-emphasis on customer relations means that service costs are driven up to a point at which the profitability and value to the contracting organisation are diminished. It may also mean high product and quality development charges to standards that are neither warranted nor expected. It may also mean giving clients over-favourable payment and charge terms – again, which are neither required nor valued.

Having said that, high levels of customer relationship are essential for the development of effective marketing activity. It is also essential that those responsible for marketing present this as a priority within their own organisations. It is very easy for those not directly concerned with marketing to lose sight of this – many executives when asked what is the most important success point on which they should focus their attention will answer: profit, efficiency, productivity, management information, sales turnover, human resource performance, technology and information systems. Identifying and satisfying customers' needs do not necessarily figure large in their list of priorities. The best organisations retain the customer at the centre of their scheme of things, supported by high-quality, capable and willing staff – and engage in productive and effective customer relations as an integral part of successful marketing.

CUSTOMER/CONSUMER/CLIENT PROTECTION

In the West, extensive legislation exists to ensure that customers, consumers and clients in every industry, commercial and public sector service get the products and services that they think they are buying when they decide to make the purchase. People who are not satisfied with what is provided have recourse to the law in order to get their money back, to have products and services replaced and also, where appropriate, to seek damages.

These disputes are expensive in financial terms and also in terms of adverse media coverage. They are also extremely bad for general morale among the staff involved. Word always gets very quickly around the industry and market sectors – and again, this is damaging to reputation and morale.

The best procedure is therefore clearly to avoid this form of dispute and to engage in practices that ensure that the interests of the customer, consumer or client are protected. Part of this means setting standards so high that mistakes and litigation become a remote possibility only. The other part means taking enough time and trouble at the outset to ensure that customers and clients know what sort of product and service they are to receive so that they choose on the basis of excellent knowledge and understanding – and this is a part of all direct sales and also more general marketing activities such as advertising and promotion. It reinforces the need for readily available after-sales activities so that any problems that do occur can be identified and resolved early and before anyone even thinks of going to law.

Moreover, most parts of the construction industry underpin their activities with codes of conduct and practice. The first point of any legal inquiry normally addresses the extent to which these codes were followed. Codes of practice vary in precise detail between sectors but their overall purpose is always to set clear guidelines of activity and performance for the quality and conduct of business relationships.

It is also necessary to take this view when dealing with lobbies and pressure groups. Increasingly, such groups are resorting to the law, especially when they do not receive proper information, fair hearings or respectful treatment; and also when specific issues have not been precisely covered in the contractor–client relationship.

STAKEHOLDERS

It is also useful to identify those groups who have interest and influence in the outcome of a proposal. These groups are called stakeholders and they are:

- **Contractor stakeholders:** shareholders; other backers; staff; customers, consumers and clients; the wider community; professional associations; consultants, both in-house and external; guarantors; professional critics; the media; shapers of trade reputation; competitors; suppliers; distributors.
- **Client stakeholders:** shareholders; other backers; political backers and factions (in the case of public projects); lobbies, pressure groups and vested interests; consultants and advisers; other opinion shapers – the media, trade press, generalists; customers; consumers; those with influence; those with expertise.

Notes

1. Clearly, some of the groups are the same. The demands of these groups differ and vary in each situation. Their influence also varies. Customers and consumers expect the contractor to do the work quickly and effectively, to restore the environment and then to withdraw – and failure on any of these counts may give a bad reputation to the facility and constitute a factor to future usage, and this loss of reputation is handed on to both contractor and client. Attitudes to the client are more directly influenced by the extent to which perceptions and anticipations are satisfied by the reality of the finished project – and this is further coloured and enhanced by any subsequent (positive and negative) media coverage and community perception.

   This continues for the duration and possibly the useful life of the project. For example, a shopping mall may have positive responses initially; but its reputation may decline if cleanliness, upkeep, security and vandalism subsequently occur (or are perceived to occur). Maintaining, servicing, upgrading and refurbishing projects are therefore a clear obligation on the part of those involved (whoever eventually is commissioned to carry them out).
2. It is essential to pin down the importance and influence of the particular stakeholders in each situation. The purpose is to identify these when

marketing, public relations, general communication and constant conversational efforts need to be applied, and the approach that is necessary in each case.

3. It is important to recognise that great influence may be exerted in spite of the level of knowledge or expertise in a particular area. Journalists and public figures pronounce on the built environment, often on no more serious a basis than they do not personally like what is proposed (or that they do personally like what is proposed). Ignorant or not, credibility is often damaged, sometimes destroyed – and sometimes unwarrantedly enhanced – and this, in turn, affects the general viability and long-term prospects of the given project – and of the contractors and clients involved.
4. The role and influence of consultants on both contractor and client sides has also to be considered. Organisations engage experts (whether independent or in-house) so that they are able to make better informed choices and decisions. The recommendations and advice of acknowledged experts normally therefore carries great weight, especially with those at the top of organisations who may nevertheless have no particular expertise in the field. They therefore lean very heavily on the recommendations of the experts that they engage.
5. Perception – public perception may be very different from what is actually proposed. Again, this is often generated by influential figures and lobbies – with or without real knowledge. It is certain that negative perceptions are enhanced by incomplete or indirect public relations (see Chapter 7). Negative perceptions have to be nipped in the bud otherwise the lasting impression is that neither contractor nor client are sure of their purpose, aims and objectives.
6. Stakeholders also have different and often conflicting legitimate interests. For example, the contractor's staff seek security of long-term employment; and this is invariably founded on the ability of the contracting firm to secure a wide variety and large volume of high-priced work. Many clients seek low-price, high-quality work, quickly delivered. Lobbies and pressure groups may set themselves against particular projects in spite of the fact that their area actually needs the employment. Public and political clients very often require monuments to their own periods of influence – and projects that may start off as very positive monuments may turn into nightmares as the result of cost overruns or the inability to produce and finish the facility to the standard that was originally envisaged (for example, as the result of budget cuts). Shareholders and guarantors require quick short-term returns on their investment; shareholders in the UK, at least, require dividends and this may affect adversely the amount of resource that the company itself requires to invest in its own future. The media often produce a good story as the result of a failure or error.

It is the responsibility of the contractor to recognise all of this and to appreciate that invariably many divergent and conflicting interests have to be reconciled as part of entering into the contracted relationship.

This is the context in which client behaviour takes place in the construction industry. Within this context however, a golden rule of marketing still applies (as it does everywhere else):

- The client buys, the contractor does not sell.
- The client buys the benefits that accrue from the product, project or service and not the product, project or service itself.

The specific benefits bought by clients in the construction industry are as follows:

- **Association:** in the use of high prestige contractors especially in the areas of design, architecture, planning, consultancy and professional services.
- **Confidence:** by engaging a particular contractor, the matter can safely be left in its hands and the result will be fully satisfactory.
- **Quality:** that once the project is completed, it will look the way envisaged; and that it will do the job or provide the facility for which it was commissioned.
- **Usage:** that the end-users – the consumers – will use the facility profitably and that it will bring real benefits to the community and environment.
- **Environment:** the finished product or facility will enhance the built environment; and will enhance the quality of life, quality of working life, communal well-being; and that it will be a general enhancement to the wider environment rather than causing blight, pollution or congestion.
- **Problem solving:** where a problem requires solution, it will be solved to the satisfaction of all concerned.
- **Priority:** the contractor will act in the client's best interests at all times and accord resources, expertise, project and service priority when necessary.
- **Value:** the cost or price agreed upon represents good value and the basis of a mutually satisfactory relationship; the price paid and facility delivered represent a good return on investment.
- **Extension:** the purchase by the client will lead to repeat business and a long-term mutually satisfactory relationship.
- **Extrinsic:** the physical benefits that accrue from the completion of the project, ownership of the product, use of the service or whatever are as previously perceived.
- **Intrinsic:** benefits of association and identity that accrue from using a particular company, product or service range that meet or exceed client expectations.

CONSULTANTS

On large and complicated building and civil engineering projects, it is usual for a range of consultants to be necessary. These may be engaged by contractor or client. Their role is to advise on any specialist aspects of the total project. This includes design (core and peripheral features; and also any secondary, tertiary or specialist features); drainage; lighting; sound; and landscaping. Other consultants are engaged to advise on specific estimates for both quantity and cost of materials and components (especially estimators and quantity surveyors). Increasingly, experts are also brought in to advise on the general management aspects of a particular project; the management of the finished project, especially facilities, environment, refurbishment and long-term consumer management experts; and again, in the provision of specialist management services.

In recent years there has been a great proliferation of consultants – with both technological and also managerial expertise. In marketing terms, the main issues that consultants have to reconcile when presenting their services to clients are as follows:

- The presentation of a distinctive high-quality professional expertise without appearing to be too specialised or compartmentalised.
- The presentation of desired qualities of flexibility, dynamism and responsiveness without appearing to be too generalist and 'all things to all people' (and therefore nothing to anyone).
- The ability to adapt both expertise and flexibility to any situation without appearing to be too bland or prescriptive.

It is also necessary to recognise current criticisms of consultancy practices (both within the construction industry and also elsewhere). The main criticisms are:

- They are too prescriptive – effectively acting as sales people for their own distinctive (and often very good) ways of doing things but without sufficient attention (real or perceived) to clients' specific requirements.
- The sheer expense of hiring top brand or exclusive consultants puts a great pressure on clients to accept their recommendations. This means that the client may often feel railroaded in a particular direction (whether or not this is indeed the case).
- People with a vested interest in a particular outcome or course of action are often perceived to hire consultants that they know will support their point of view.
- Using consultants to produce an independent point of view is often seen within the contracting client or hiring organisation to give an entirely spurious rationale to what may otherwise be a contentious plan of action, or controversial project.

Consultants from all disciplines have to be aware of these criticisms. Those who hire consultants are extremely aware of them. The key to effectiveness lies in the ability to harmonise expertise, flexibility, exclusivity and presentation in such a way as to ensure that the client is getting the best aspects of each; and that the result of hiring the consultant (from whatever discipline) is either to increase the value of the project or to reduce its cost.

The main difference between these benefits in the construction industry and in other sectors is the need to pay attention to the wider picture. Each of these has to be seen in terms of the ultimate end-users as well as the client. They in turn will have their own perceptions of each of the points above – and as already noted, these may not accord with the client perceptions.

CLIENT EMPATHY

Seeing the client's needs from the point of view indicated above is the first step towards empathy or identifying closely with the clients and their purposes. Beyond this, it is necessary to recognise where the proposed contract occurs in the client's 'hierarchy of needs'.

Hierarchies of needs

This means considering and placing the product, project or proposal in the following ways (see Figure 2.1 and also Summary Box 2.1):

- **Basic or fundamental needs:** extreme pressure on clients to produce basic facilities such as housing, roads, other infrastructure; sometimes this extends to commissioning industrial, commercial and public development with the view to providing work for the locality.

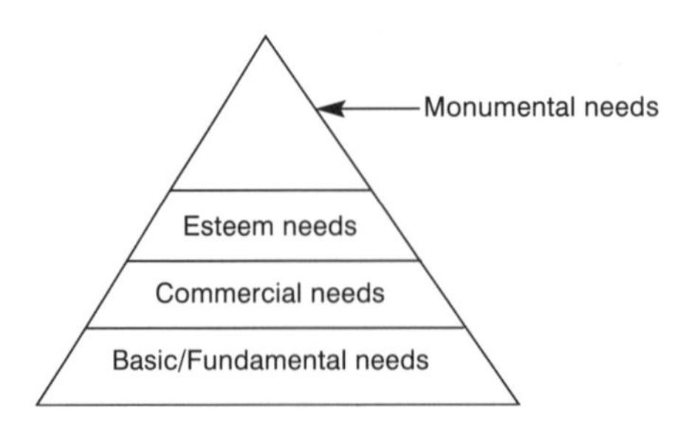

Figure 2.1 Hierarchy of needs adapted: an indication of likely and possible sources of work, and client perspective.

- **Commercial needs:** the provision of facilities and amenities that in turn provide the client with long-term commercial returns.
- **Esteem needs:** the need to identify with, and be identified with, the commissioning of prestigious public and commercial works.
- **Monumental needs:** the need to provide long-term, eternal, enduring monuments of some sort. In the past, these were cathedrals, castles and city walls. At present, and for the foreseeable future, these still include some religious buildings; and also industrial, commercial and domestic facilities (e.g. the Barbican Centre in London; the regeneration of the London Docklands; the refurbishment and rebuilding of the Royal Docks at Liverpool).
- **Client needs:** these are security and confidence in all situations. Whatever the product, project, service or expertise applied, the client's need for a basic level of comfort in dealings with the contractor must be addressed. This then becomes the basis for establishing the precise nature of needs in given circumstances – but without this as the basis, no further progress is possible.

In marketing the construction industry two further pressures are apparent: the cost of the contract; and the quality of the contract. In this context, these are the foundations that underpin security and confidence – once the relationship is entered into, the client is entitled to expect that the work will be delivered to the required cost and required quality; and usually within the agreed timescale.

SUMMARY BOX 2.1 Contrasting Views of the Customer Position

Stew Leonard

Stew Leonard is a highly profitable supermarket on the edge of a small town on the eastern seaboard of the United States. The company has been studied extensively during the 'Excellence' investigations of the late 1970s and 1980s. As far as the customer position is concerned, the supermarket has two rules:

1. The customer is always right.
2. If the customer is wrong, see rule 1.

The customer/client is always right

In a speech in 1986 Alan Sugar, the Chairman and Chief Executive of Amstrad, described this statement as 'rubbish'. It certainly has to be seen in context. Clients may not know precisely what they do want but expect this to become clear in preliminary discussions with prospective contractors. The client's representative may have been given a brief (for example, by a board of directors or political masters) that they know to be unworkable – and are expecting all prospective contractors to tell them this. On the other hand, clients certainly do not want the reality or the enduring perception that they have had their arms twisted by the contractor to adopt a particular approach, to go in a certain direction or choose a particular project contrary to their own wishes.

The introductory problem
This occurs especially in the planning, consultancy and design sectors. It occurs when the contractor is invited in ostensibly to address one problem, when the reality is different and often far more serious and fundamental. Effective marketing here means identifying this at an early stage of the discussions and, if necessary, leading the client into the broader argument (though very often this will not be necessary, rather the client will be extremely grateful and suitably impressed that the prospective contractor has latched on so quickly).

The politicians' monument
This is a major concern in the run up to the year 2000 – the Millennium. Politicians of all nations – including the UK – are seeking prestigious projects and schemes of a distinctive design and unique features as lasting monuments to their own brilliance. This is of greatest concern to high prestige, high reputation and highly expert design, planning and building consultants, contractors and architects for national and international mega-projects – where they are going to receive the commissions to design and build these monuments; and where they have to be completed at the latest by 31 December 1999 (the fact that the Millennium genuinely does not start until the year 2001 is not an issue in this context!). They are also inevitably going to have to put up with political interference – and the unilateral withdrawal of backing if such projects suddenly become unviable, or if the political mood changes, or if the particular politician who backed the initiative falls from grace.

The problem is the same in some locations for smaller operators. Problems occur – and are starting to occur in Millennium projects – where an initially brilliant idea runs out of steam, or is downgraded, or is superseded by something even more brilliant. Problems also occur where political funding and priority changes occur; and where the particular politician or over-mighty figure resigns, moves on, is disgraced – or where the political influence changes and replaces the previous brainwave with its own.

CONVENIENCE, CHOICE AND SPECIALISM

Customers and clients in every sector expect a combination of convenience, choice and specialism when making their purchases. Whatever the sector, the following generally holds good: convenience is a key factor once general confidence in the product or contractor has been established; where this confidence has not been established, customers and clients go to a lot of trouble – and inconvenience – to seek sources of satisfaction.

Everyone expects a measure of choice, whether they are buying for themselves or on behalf of an organisation; they do not like to be told that there is no choice – whether or not that is the case.

The purchasing of specialist products and services are those for which a significant effort is needed or given; clients and customers therefore tend to put themselves out when they want a very precise product or service; and the greater the accessibility and deliverability of this on the part of the contractor, the less of an effort required by the customers or clients (see Summary Box 2.2).

The relationship between convenience, choice and specialism has to be reconciled in the presentation of construction products and services. Problems with this arise from:

- Time, resource and other pressures on the client.
- The expertise of the client which is: at least on a par with contractors at the prequalification stage; at best on a par with contractors at any presentation to client boards of directors; very unlikely to be on a par with contractors in the presentations to boards of governors and other political elements; extremely unlikely to be on a par with the contractors in presentations to lobbies and vested interest groups.
- Other behavioural factors: key features are where the client does not have the contractor's level of expertise. Choice is therefore certain to be made on a combination of the client's perception of the contractor's expertise, together with personal empathy, perceived understanding, personal like/dislike – and a return to the buyer behaviour with which they are most familiar, the combination of convenience, choice and specialism that they themselves use when making individual purchases.

PERCEPTION

Understanding perception is a major step in becoming an effective marketer. Perception is the term used to describe the ways in which different people and groups see and understand their environment and the ways in which they use, limit and interpret the information and impressions with which they are presented. This information can then be transformed into something that is useful and usable. Part of the process is based on the senses – sight, hearing, touch, taste and smell. Part of it is based on instinct – for example, one's view of what is edible is clearly coloured by how hungry one is. A part of it is learned – and this comes from a combination of civilisation and socialisation, which, in this context, gives rise to the organisational ways in which behaviour is regulated and attitudes are formed. It is the basis of the formation of norms, expectations, customs, etiquette and *realpolitik*; and forms the basis for concepts of

SUMMARY BOX 2.2 Buying Behaviour: Planning Committees

In 1984, a local government planning committee had two items on its agenda – the siting of roads and pedestrian signs to a public library in one of the towns in the district; and the commissioning of a major medium to high quality housing development in another of its small towns. Two hours were allocated for the purpose. It took the two items in this order:

1. *The signs to the library*
   These were considered from every possible point of view. The number of signs; whether to locate them on the left or right of the approach roads (or on both sides); the colour to be used on the signs; whether they would be equally convenient for drivers, pedestrians, tall, short and disabled people; whether to subcontract the work or whether the in-house contractor would have the resources to do the work – all were considered and debated at length. Finally, seven signs in specific locations were determined and the work given to the in-house contractor.
2. *The housing development*
   This was a proposal for 30 detached and semi-detached houses with a combined value at the time of £2.5 million. Less than ten minutes was available at the end of the meeting to discuss and approve this. By chance, one of the members present had thought to ask a potential contractor for some ideas and had brought along an outline sketch produced by the company as the result. This was approved as the planning committee members were putting on their coats to go home.

There is also another aspect which has, from time to time, influenced county council committee meetings in the past. Large constructional projects are passed quickly as many lay people cannot decipher the plans or comprehend such vast sums of money. At the other extreme, lengthy debates ensue on the award of contracts for the supply of small items, such as dusters, as everyone is familiar with them.

This may not represent a typical planning committee meeting. However, it is an extremely useful indication of the importance of considering the behavioural aspects of buying and purchasing.

fashion and desirability, and confidence in doing business with others. There are also more general situational factors such as peer group pressure, value judgement, media pressure and business activity. And marketing is directed in large part to the formulation of positive and acceptable impressions.

Comfort and liking

The initial basis for confidence between customer and client is based on a positive initial meeting point. Comfort and liking occur when elements and features accord and harmonise with each other. Instant rapport is achieved when initial perceptions coincide, meet expectations and lead to an initially productive relationship. This is developed as people become more familiar with, and knowledgeable about, each other and about situations and circumstances.

The greater the continuing coincidence, the greater the harmony and accord, and the more flexible the boundaries of this become.

Discomfort and dislike occur when the elements are in discord. This is usually founded in strong and/or contradictory initial and continuing impressions. For example, a high-quality and beautifully presented contractor specification may arrive on the client's desk; a meeting is arranged with a contractor's representative; and the contractor's representative turns out only to have a very general knowledge of the broader situation and no understanding of the client's needs. Or it may be as simple as the fact that the contractor's and client's representatives can simply find no personal positive meeting point. To the unwary, this simply becomes the instant basis for a non-productive relationship. In practice, this may turn out to be the case; however, the initial limitations do first have to be accommodated (see Summary Box 2.3).

Inference

People infer or make assumptions about others and about things, situations and circumstances based on the information available and their interpretation and analysis of it.

For example, it is not possible to define attitude from behaviour or performance from attitude; it is only possible to infer this. This means that while it is possible to predict to a certain extent, there are too many uncertainties and the full range of actual results remains possible and available until the situation is developed further. For example, a contractor's representative may have an excellent personal and professional relationship with his/her opposite number at the client organisation; to the unwary the inference is that this will 'inevitably' or 'certainly' lead to good volumes of work. Great disappointment is experienced therefore whenever this does not happen. Clearly, therefore, the relationship has to be seen in broader terms – part of the composition of a 'strong and positive personal and professional relationship' must include the delivery of results as well as the comfort of the interaction.

Extreme forms of inference are jumping to conclusions and gut reactions – in which quantum leaps are made about the outcome of something from a limited range of information available. In each case, individuals select those elements before them with which they are familiar, and place

SUMMARY BOX 2.3 **Perceptual Errors and Limitations**

The sources of error and limitation are as follows:

- not collecting enough information;
- assuming that enough information has been collected;
- not collecting the right information; collecting the wrong information;
- assuming that the right information has been collected;
- seeing what we want and expect to see; fitting reality to our view of the world;
- looking for in others what we value for and in ourselves;
- assuming that the past was always good when making judgments for the future;
- failing to acknowledge other points of view;
- failure to consider situations and people from the widest possible point of view;
- unrealistic expectations, levels of comfort and satisfaction;
- confusing the unusual and unexpected with the impossible.

The remedies are:

- understanding the limitations of personal knowledge and perception; that this is imperfect and that there are gaps;
- deciding in advance what knowledge is required of people and situations, and to set out to collect it from this standpoint;
- structure activities where the gathering of information is important – this should apply to all marketing research, questioning, product development, project prequalification activities, sales pitches and sales staff;
- avoid instant judgments about people and organisations however strong and positive or weak and negative the first impression may be;
- building expectations on knowledge and understanding rather than instant impressions;
- ensure exchanges and availability of good quality information;
- ensure open relationships that encourage discussion and debate and generate high levels of understanding and knowledge exchange;
- train and develop self-awareness and understanding among all staff, especially marketing staff;
- recognise and understand the nature of prevailing attitudes, values, beliefs and prejudices;
- recognise and understand other strong prevailing influences – especially language, nationalist, culture and experience.

their own interpretation of the likely or 'logical' outcome based on their knowledge and understanding of what happened before when the elements were present. The range of misunderstandings possible is virtually limitless. There may be three elements out of six present with which the individual is familiar; or this may be three out of ten; three out of forty;

or three out of a hundred. In each case, it is the familiar three that form the basis of judgment.

Halo effect

This is the process by which a person is ascribed a great range of capabilities and expertise as the result of one initial impression of an overwhelming characteristic. The person who has a firm handshake is deemed to be decisive. A person with a public school education is deemed to be officer material. The person who can play golf is deemed to be an expert in business.

The converse of this is the 'horns effect' (the halo apparently comes from heaven; the horns therefore clearly originate elsewhere). This occurs when a negative connotation is put on someone or something as the result of one (supposedly) negative characteristic. Thus the person with the soft handshake and lisp is deemed to be soft and indecisive. Anyone who wears fashions from a past era is deemed to be eccentric or old fashioned.

Stereotyping, pigeonholing and compartmentalisation

It is a short step from the halo effect to develop a process of stereotyping, pigeonholing and compartmentalisation. This occurs at places of work whereby, because of a past range of activities, somebody is deemed to be 'that kind of a person' for the future. This may both enhance and limit careers, activities and organisation progress dependent on the nature of the compartmentalisation. In any case, it gives specific and limited direction. The contracting company that has worked for a long time with public service organisations is likely to get pigeonholed by its sector as 'a public service contractor'; if it wishes to work elsewhere, it has to get over this broader perception as well as convincing future clients of the extent of its capabilities.

Self-fulfilling prophecy

Self-fulfilling prophecy occurs when a judgment is made about someone or something. The person making the judgment then singles out further characteristics or attributes to support this view and edits out those that do not fit in.

Perception mythology

This occurs where myths are created by people as part of their own processes of limiting and understanding particular situations. A form of rationale emerges which is usually spurious. Thus for example, people will say such things as:

- 'I can tell as soon as someone walks in the door whether he can do this job'; or
- I always ask this question of potential contractors and it never fails'; or
- 'I never trust people in white shoes/white socks/with moustaches/ with tinted glasses';

in order to give themselves some chance of understanding and therefore mapping the person who stands before them.

People also use phrases such as 'in my opinion' and 'in my experience' for the same general reasons and also in support of general views of the world's stereotypes and business developments.

People also relate the present situation to something that happened in the past or elsewhere. The line of argument is 'X did this and it worked so we should do it and it will work for us'; or 'it happened liked this in 1929 and so this is the way to do it in 1998'; or, more insidiously, 'we had to do things like this in my day and it never did us any harm and so this is how it has got to be done now'.

It is a short step from this to comparisons, again often of dubious value. Industries publish league tables of company performance by business volume, business value, wage and salary levels, numbers employed, variety of locations and so on. These are then used to justify and explain a range of other issues of varying degrees of relevance and substance. Companies, for example, say 'we are 6th in the league' or 'we are no worse than anyone else in the sector' or 'we are doing alright given the state of the industry at large' – without attaching any rationale to any of the points made; and without any reference to the fact that the company must stand on its own in the end. Different parts of the construction industry publish these league tables without any real explanation or understanding of what they mean; for example, a civil engineering company that comes top of the sector for having work worth £10 million in a given month/quarter normally gives no indication as to whether this £10 million is for one contract or whether it is ten contracts at £1 million or a hundred contracts at £100,000.

Mapping and constructs

People, situations, activities, images and impressions are being fitted into the perceived map of the world in ways which can be understood, managed and accommodated. Any information gathered is broken down into 'constructs' – characteristics that are categorised as follows:

- **Physical:** by which we assume or infer the qualities of the person from their appearance, racial group, beauty, style, dress and other visual images.
- **Behavioural:** in which we place people according to the way they act and behave or the ways in which we think they will act and behave.
- **Role:** whereby we make assumptions about people because of the variety of roles that they assume; the different situations in which they assume these roles; their dominant role or roles; and the trappings that go with them.
- **Psychological:** whereby certain occupations, appearances, manifestations, presentations and images are assumed to be of a higher order of things than others. This reflects the morality, values and ethics of the society of the day, as well as the environment and organisation in question.

This part of perception aims to build up a picture of the world with which the individual can then be comfortable. Comfort is achieved when people and situations are perceived to have complementary characteristics or constructs. This comfort occurs when characteristics and constructs are contradictory. For example, there is no difficulty placing an individual who is kind and gentle on the perceptual map; there is difficulty however, in being comfortable with an individual who is both kind and violent (see Summary Box 2.4).

First impressions

First impressions count – this is received wisdom throughout the world. It applies everywhere. Yet first impressions are plainly misleading: *prima facie* we must know less about someone after 30 seconds than after 30 minutes or 30 years. Yet, overwhelmingly, the converse is highly influential and this should be understood.

The first impression gives a frame of reference to the receiver. The appearance, manner, handshake and initial transaction are the writing on a blank sheet. Before there was nothing; now there is something on which to place, measure and assess the other. The impact is therefore very strong. It is essential to recognise this. The consequence of not doing so is that a one-dimensional view of the individual or organisation is formed. Everything that is contrary to that dimension, or that

SUMMARY BOX 2.4 **Perception: Other Factors**

The other elements of which those involved in marketing should be aware are as follows:

- **Closure:** which occurs where an individual sees part of a picture and then completes the rest of it in the mind; or hears part of a statement or conversation and then mentally finishes it off.
- **Proximity:** based on the desire to understand and be comfortable with that which is close at hand to be at ease in the immediate environment. Matters that cause discomfort therefore assume importance out of proportion to their actual effect.
- **Intensity:** the effects of extremes of heat, cold, light, darkness, noise, silence, colour, taste, touch, smell, comfort and discomfort.
- **Attribution:** the explanation put by individuals on behaviour or activities; especially when trying to explain a quirk or inconsistency.
- **Confidence:** people who have complete confidence in each other speak in direct language; people who do not must couch what they say in safe phrases to avoid giving offence.
- **The messenger:** messages from one quarter may be unacceptable; the same message delivered by someone else may be eagerly accepted.
- **Language:** people perceive situations partially according to the language used; and according to the language used, the same situation may be seen as an opportunity, challenge or chance for progress, or as a threat, risk or uncertainty. Moreover, the more direct, clear and unambiguous the language used, the greater the general and positive response generated.
- **Repetition:** the repetition of message gives currency, familiarity and validity; repetition must therefore be consistent.
- **Authority, responsibility and position:** influencing the propensity of the receiver to accept or reject a particular message. A person who understands well the strength or weakness of his/her own position is always likely to be a more effective communicator than one who does not.
- **Visibility:** a prerequisite to effective communication. Visibility greatly helps in the generation of confidence, familiarity and interaction. Lack of visibility is both a perceptual barrier in itself and also compounds any others that may be present. Misunderstandings occur least often and are most quickly and easily sorted out at face-to-face level.

indicates complexities, other dimensions and qualities is edited out (see Table 2.1).

This table is a useful (but by no means perfect or complete) means of compartmentalising the cues and signals that are present when coming into any situation or into contact with someone or an organisation for the first time. There are certain to be contradictions. It is essential to recognise and understand this in order, in turn, to understand the impact and influence of first impressions.

Table 2.1 *First impressions.*

People	*Service*
Appearance, dress, hair, handshake Voice, eye contact Scent, smell Disposition – positive, negative, smiling, frowning Establishing common interest/ failure to do so Courtesy, manner Age	Friendliness/lack of Effectiveness Speed Quality Confidence Value Respect Ambience Appearance
Objects	*Organisations*
Design Colour, colours Weight Size Shape Materials Purpose, usage Price, value, cost	Ambience Welcome Appearance Image and impression Technology Care Respect for others Confidence Trust

Defence mechanisms

People build defences (blocks or refusal to recognise) against people or situations that for some reason are unacceptable, unrecognisable, threatening or incapable of assimilation. Perceptual defence normally takes one or more of the following forms:

- **Denial:** refusal to recognise the evidence of the senses.
- **Modification and distortion:** accommodating disparate elements in ways that reinforce the comfort of the individual.
- **Recognition but refusal to change:** where people are not prepared to have their view of the world disrupted by a single factor or example. This is often apparent when people define 'the exception to the rule'.
- **Outlets:** where the individual seeks an outlet, especially for frustration or anger away from its cause – for example, brow-beating a subordinate or blaming a junior offers a sense of relief to someone who has failed to get a contract or who has previously been brow-beaten themselves by a superior.
- **Recognition thresholds:** the higher the contentiousness or emotional content of information, the higher the threshold for recognition (that is, the less likely it is to be readily perceived).

- **Adaptation:** our view of the world is influenced directly by the circumstances and surroundings in which we find ourselves; and this relates to priority levels – what is important *now*; and this changes as situations change and progress is made. This is often apparent where, for example, an organisation fails to win a piece of work – the organisation refuses to see why it has failed and how it can adapt itself for the future.

ATTITUDES

Attitudes are the mental, professional, moral and ethical dispositions adopted by people to others and the situations and environments in which they find themselves. They can be broken down as follows:

- **Emotional:** feelings of positiveness, negativeness, neutrality or indifference; anger, love, hatred, desire, rejection, envy and jealousy; satisfaction and dissatisfaction. Emotional aspects are present in all work as part of the content, working relationships, reactions to the environment and the demands placed on particular occupations.
- **Informational:** the nature and quality of the information present and the importance that is given to it.
- **Behavioural:** the tendency to act in particular ways in given situations. This leads to the formation of attitudes where the behaviour required can be demonstrated as important or valuable; and to negative attitudes where the behaviour required is seen as futile or unimportant.
- **Past experience:** memories of what happened in the past affect current and future feelings.
- **Specific influences:** especially those of peer groups, professional groups, work groups and key individuals such as managers, supervisors, contractor representatives. These also include family and social groups and in some cases religious and political influences.
- **Defence:** once formed, attitudes and values are internalised and become a part of the individual. Any challenge to them is invariably viewed as a more general threat to the comfort of the individual.

VALUES

Values are the absolute standards by which people order their lives. Everyone needs to be aware of their own personal and professional values so that they may deal pragmatically with any situation. This may extend to marked differences between individuals or between an individual and demands of the organisation or between the representatives of contractors and clients. Conflicts of value often arise at places of work; anything to which people are required to ascribe must recognise this, and if it is to be effective, must be capable of harmonisation with the values of the individual. These values may be summarised as follows:

- **Theoretical:** where everything is ordered, factual and in place.
- **Economic:** making the best practical use of resources; results orientation; the cornerstone of people's standards and costs of living.
- **Aesthetic:** the process of seeing and perceiving beauty; relating that which is positive and desirable or negative and undesirable.
- **Social:** the sharing of emotions with other people.
- **Integrity:** matters of loyalty, honesty, openness, trust, honour and decency; concern for the truth.
- **Political:** the ways and choices concerning the ordering of society and its subsections and strata; the ways and choices concerning the ordering of organisations.
- **Religious and ethical:** the dignity of human kind; the inherent worth of people; the absolute standards of human conduct.
- **Professional:** carrying out professional and organisational activities; accepting recognised and agreed obligations; conducting activities in ways with which the individual is comfortable; working within standards, codes of practice and legal constraints.
- **Shared values:** the ability to harmonise different values – contractor–client; contractor–client representatives; client–consultant; contractor–consultant; contractor–client–consumer–user; work commissioners; industrial and sectoral standard setters; pressure groups and vested interests; rules and regulations; absolute standards of conduct.

Training in the understanding of perceptions, attitudes and values is a key feature of the best marketing development programmes. It is the key to building positive and effective working relationships and developing opportunities for profitable activity. Especially at the civil engineering and contracting end of the industry, the formation of long-term, enduring and consistent attitudes and perceptions is certain to help to give a sustainable competitive advantage in contractor–client relationships.

The point that is being addressed here is the difference between:

- 'Would you like to use us?' and
- 'Will you use us (and if so, when, what for and how often)?'

'Would you?' is a general comfortable question. It binds nobody to anything. The answer, all things being equal, is therefore invariably 'yes' – and if this is confused with 'will you?' gives a dangerous over-estimation of the prospect of work. This is known as 'the generally favourable response' and is *not* to be confused with 'will you?' to which the answer is either yes, no or maybe. The result of getting to this point is that people have a much clearer indication of where they stand at an early stage in the potential relationship and can therefore structure and schedule things accordingly.

Marketing effort is required that has the purpose of getting potential clients from 'would you?' to 'will you?' – thus turning them into actual clients. Much of this is based on developing the perceptions, attitudes and values, and recognising and understanding the constraints indicated earlier.

This is true at all levels across the industry and in all its activities. A domestic based multinational corporation may tender for major projects on the grounds that it can do them and that it is well known in general to the potential client – and yet fail to get them, or may be even losing out to foreign competition. A local or regional builder may have an excellent image and reputation in its own area, and yet again not gain all the contracts expected. A sole trader jobbing builder may be known, liked and respected for the quality of his/her work, and yet still lose out to others less well regarded. In each of these cases, the brief description given can be summarised as a generally favourable response rather than as a fundamental cornerstone for a profitable working relationship.

The converse is also true. The answer to 'would you?' may be 'no' – and yet the answer to 'will you?' when it is finally asked in this way may nevertheless be 'yes'. This may be brought about by a lack of choice at the precise point when the work needs doing or by the fact that the contractor in question can start and finish the work more quickly than others or because it offers the lowest price and the client has to take the lowest price (as, for example, in public service contracting). The opportunity therefore arises to do the work and from that, to build a direct, specifically favourable, response and to change attitudes as well as, and alongside, generating repeat work. For a view of this converse, see Summary Box 2.5.

Understanding customers, consumers and clients therefore means understanding human and professional behaviour, and the constraints placed upon it. The chief concern is recognising what is important and essential to clients and client's representatives in particular situations, and presenting the capabilities of the contracting or potential contracting organisation in ways that meet this to best advantage. Clearly, therefore, the rational approach is not enough – it is no use having the best products, technology or expertise if these cannot be presented in ways that meet customer and client needs. And it is also not enough simply to say 'we can and will meet your needs'! Time, trouble, energy and resources have to be invested, in finding out what those needs genuinely are, learning and understanding the particular situation and, from this, generating effective marketing activity and forms of presentation that meet the human and professional aspects.

SUMMARY BOX 2.5 **Attitudes: Japanisation**

Japanese industry has had a major influence on all aspects of consumer society and Western civilisation over the past thirty years and this is set to continue. Japanese tourists account for a significant percentage of the global travel and tourist trade. Japanese manufacturing has transformed the standards, quality and performance of cars, shipping and domestic electrical goods and made these goods more widely accessible and available than ever before. Japanese construction and civil engineering practice has transformed the ways in which contracts are carried out in the Far East and United States and is increasing its influence in Europe and the UK.

Prevailing attitudes and prejudices towards Japan and its industry remain mixed. Globally, there is concern about the economic conquest and dominance achieved in these fields, and the consequent dependence of consumers on Japanese organisations. Organisations from elsewhere have had to adapt or die – and in many cases it has been the latter. And individuals from many parts of the world who have fought in wars against the Japanese, often suffering greatly as a consequence, nevertheless buy and use the Japanese products in preference to those made elsewhere.

In the UK, the domestic construction industry has planned, designed, built, fitted out and maintained the facilities for incoming Japanese banks, factories, schools, social and leisure facilities. And again, this is not because of a generally favourable attitude. It has rather been driven by the hard side – that the Japanese companies were willing to do rather than discuss; that they were willing to invest and commission work; that, provided the quality was right, they were willing to pay well; and that, provided the finished facility was delivered to time, quality and cost specifications, they were willing to provide repeat business.

3 Strategy

Like all business activities, marketing needs a strategic approach – a clear, long-term and sustainable direction as the basis for activities. This relates directly to organisations' overall chosen strategies and directions because marketing is the prime activity that is to translate what is envisaged into effective and profitable action. The only remote exception to this is where a company has a long list of work extending into the far distant future – and even then, clients may and do go elsewhere if new players come into the sector with their own version of capabilities and expertise that gives them a sustainable advantage.

Lack of marketing is normally a sign of complacency or uncertainty, and this is often borne of having had a long and assured series of contracts from similar or familiar sources on which considerable levels of prosperity have been founded. In general, there is now no such thing as a protected sector or assured source of work. Clients have choice. Marketing is therefore essential and for best effect this is designed to have a direct and positive purpose – of strategy or direction.

The purpose of this chapter is to go through the different stages by which successful and effective marketing strategies are established. This means:

- Analysing the environment, searching for opportunities, niches and segments capable of exploitation.
- Analysing the markets and segments that are to be concentrated on; analysing the actual and potential customer bases and establishing their needs, wants and expectations.
- Identifying and establishing a clear strategic standpoint or position.
- Developing this into profitable, effective and successful activities.
- Developing a long-term future.

ANALYSING THE ENVIRONMENT

The purpose of analysing the environment is to match external conditions and trends with organisational capacities, capability and willingness to operate in the chosen area. The basis for doing this is given in Figure 3.1.

This approach is then further developed as shown in Figure 3.2 and Table 3.1.

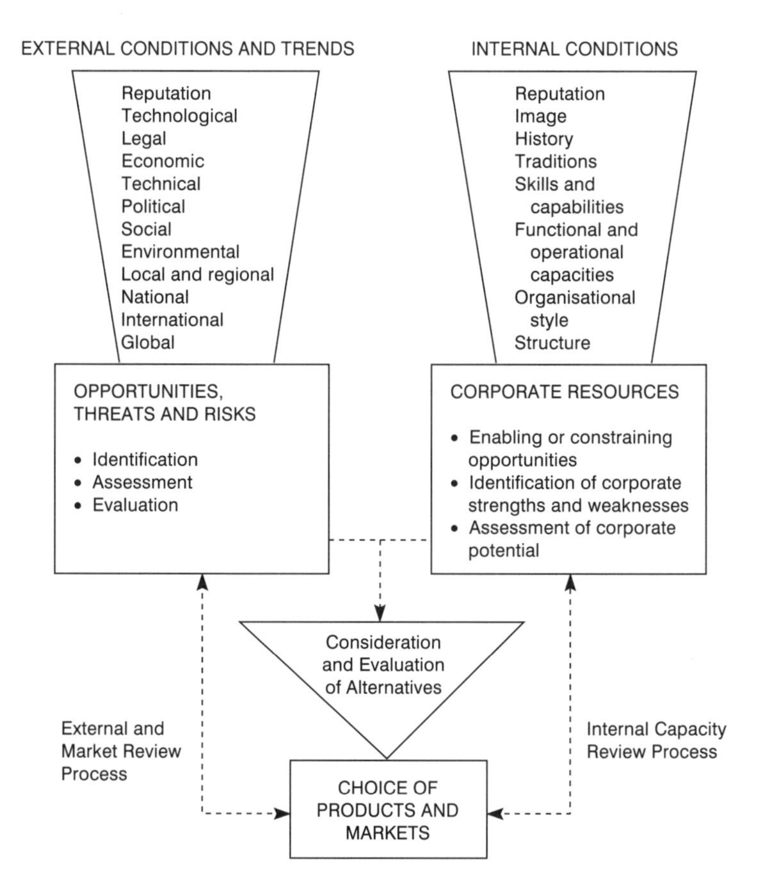

Figure 3.1 Matching opportunities with capability.

Table 3.1 *Indicators of organisation strength and business potential.*

Indicators of organisation strength	*Indicators of business potential*
Market share	Market size
Marketing activities	Market state: growth, maturity, decline
Customer service	Market potential
Confidence and image	Cyclical and seasonal factors
Nature of operations	Market structure
Financial structure and strength	Nature of competition
Product range and potential	Nature of profitability
Product strengths, quality and reliability	Entry and exit barriers
Managerial quality and expertise	Legal factors
Technological quality and expertise	Social and cultural factors
Staff quality and expertise	Ethical factors
Managerial and staff willingness	Specific drives and constraints

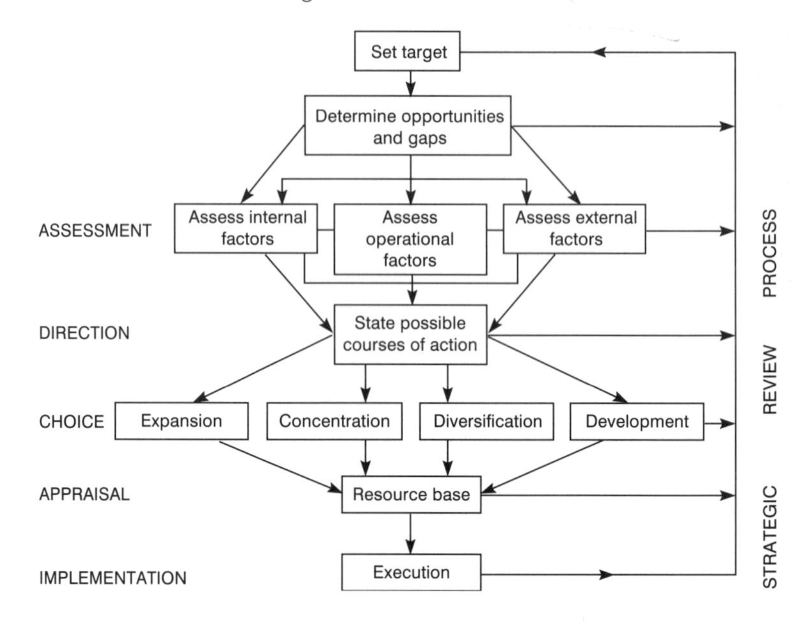

Figure 3.2 Indicators of organisation strength and business potential.

The outcome of this should be a thorough general understanding of the markets, segments and niches in which activities are being considered. To be successful, considerable resource investment is necessary – at this stage, organisations need to have a clear picture of likely or potential chances of success or failure and the main factors that have to be considered.

Once the decision has been taken to pursue matters further, much more detailed analysis of the environment is required. A detailed examination is conducted in much more depth of the given areas of activity. A familiar approach to this is the use of Porter's 'five forces' model, as shown in Figure 3.3.

The outcome is a detailed understanding of the ways in which the particular sector behaves and operates. It makes distinct reference to five specific areas:

- **Rivalry:** the current nature of competition between those conducting their business and competing at present.
- **Threat of entry:** the potential effects of new players into the sector on existing activities; the extent and prevalence of potential entrants.
- **Substitutes:** choices and alternatives that the client base could make away from the current range of activities; alternative routes to the same destination; extending the range of choice available to clients.
- **Backwards integration:** in which the potential for existing players to strengthen their position by buying up sources of supply and components is examined.

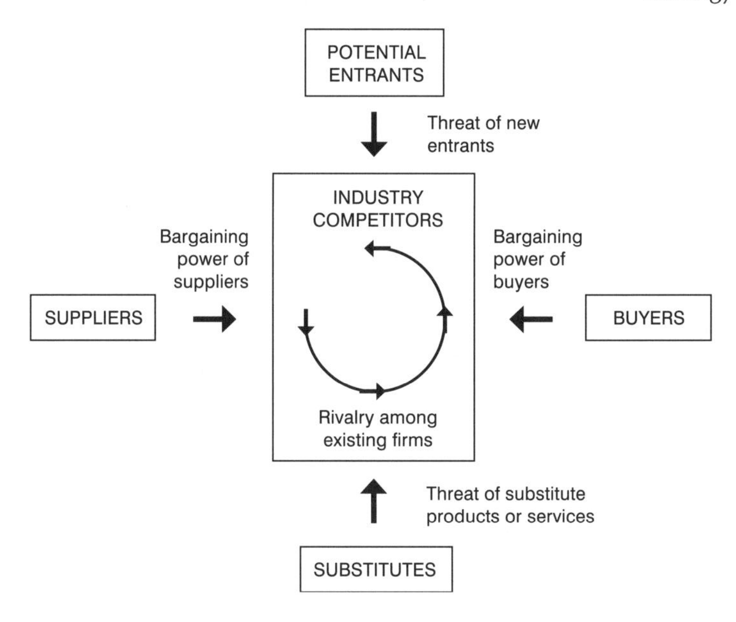

Figure 3.3 Michael E. Porter: 'Five forces' model of competitive and marketing strategy.

- **Forwards integration:** in which the ability of existing players to dominate the means of distribution is considered.

This information can then be used to pinpoint specific issues.

Entry barriers

The main barriers to entry are:

- The existing strength of current operators and the propensity of the client base to continue to use them.
- The availability of the necessary technology and the expertise to use it.
- The extent and nature of capital investment required to gain a foothold and a competitive position.
- Any other cost barriers that may be incurred – especially 'switching costs' which are incurred when an organisation decides to move from one sphere of operations to another.
- Familiarity and behavioural barriers – which refer to the willingness of the sector to take on or consider new operators in preference to those already present; and which refer also to the willingness of the organisation, its staff and other stakeholders to consider itself a serious player in a new sector.

- Legislation – the extent to which the sector is protected by specific legal boundaries and any specific steps necessary to work within these.
- The ability to differentiate, in order to be able to offer a real or perceived distinctive advantage to the client base in preference to existing players.

Exit barriers

For the new or potential entrant, the costs and consequences of failure have to be assessed in advance (see Summary Box 3.1). The main exit barriers are:

- Costs and charges that will, or may, have to be borne if the activities do not match up to their perceived or forecast potential.
- Confidence – especially the adverse affects in confidence on staff and clients if withdrawal does become necessary.
- Reputation – again whether this would be adversely affected as the result of having to withdraw.
- Destabilisation – whether the withdrawal from one sector is likely to lead to destabilisation, either within that sector or within the organisation itself.
- Wastage and losses that have to be borne as the result of hiring technology and expertise that subsequently are no longer needed.
- Excess and spare capacity that may have to be sustained in the short and medium-term, at least if withdrawal leaves a part of the organisation under-employed.

SUMMARY BOX 3.1 **Risk Analysis**

The information gained from using the 'five forces' approach may also be used to study and evaluate risk. The purpose of risk analysis is to measure and evaluate the possible range of outcomes that may be expected to be achieved – and therefore the extent of risk incurred in going into a particular sector or range of activity. This is distinct from uncertainty – which is the outcome of unplanned and unstructured activities.

The components of risk analysis are as follows:

- market and sectoral trends – whether they are growing or declining; static, stable or stagnant;
- assessment of substitutes and alternatives and factors that may lead the client base to consider these;
- social, political, legal, economic, ethical and environmental considerations;
- operational aspects, including levels of resources, staffing, skills and capabilities; and projections of these for the future;
- organisation constitution, style and attitudes;
- critical success requirements – both in hard terms of profit turnover, sales and activity volumes; and also the soft elements of reputation, ethics and standing;
- full consideration of the best, medium and worst possible outcomes of the proposed range of activities;
- factors outside the control of the organisation and its managers;
- crises and early warning systems; and processes for crisis and early warning management;
- the ability to support these with quantitative, statistical, financial and mathematical approaches;
- the view of influential stakeholders, pressure groups and lobbies; the organisation's political systems and the extent to which they are able to accommodate or block the proposed range of activities;
- decision-making processes, chains of command and channels of communication;
- the prevailing culture values, ethics, attitudes and beliefs of the organisation.

The purpose is to arrive at an assessment of risk; the ability to conduct risk-free activities is not possible. It is the basis for informing strategic and managerial judgment. It needs to be applied to all aspects of activities: the organisation and its capabilities, the markets served and their properties, potential markets, the wider environment and the community at large.

Rivalry

This refers to the real nature of competitive activity currently being conducted in the sector. It means considering the marketing processes and activities currently present and the extent to which these are familiar and acceptable to the client base. It is also necessary to consider the industry structure, e.g. whether the sector is dominated by one or two key players, the size of market share commanded by the main or major players, and whether it is necessary to take business from existing players in order to be successful, or whether the market can be expanded by new entrants. It is also necessary to consider the nature and strength of marketing activities and the extent to which these can be varied (a) to suit the client base, and (b) because of the potential entrant's current marketing position and activities.

Substitution

This means assessing the sector in terms of the alternative uses that it could or might make of its resources. This involves considering the relative price of substitutes, their relative availability and what may trigger the client base to use them. It also means consideration of new ways of carrying out existing activities and using new materials to produce existing equivalent products.

Command of resources

This refers to the relative ability to gain access to the technology, supplies, components and primary resources necessary to operate effectively in the sector. In some activities, this is no problem – primary resources and generic technology are readily available. In others, it is necessary to have specific direct access to these resources – and the problem may be compounded by the fact that existing players have already secured preferential agreements with the suppliers or indeed, bought up the sources of supply themselves.

Distribution

This refers to the ability to get the product or service to market, either directly or through the use of intermediaries. Again, it becomes a problem when the means of distribution are not easily available to the potential entrant. It may also be a problem when it is necessary to use agencies that have continuing obligations to existing players; or when the means of distribution and outlet have been bought up or dominated by existing players.

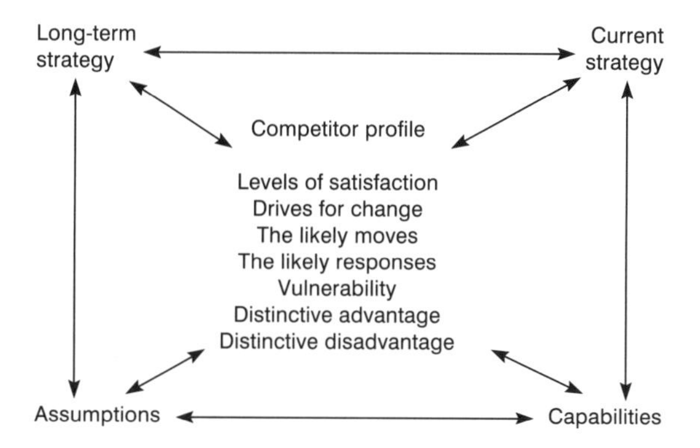

Figure 3.4 The components of a competitor analysis.

COMPETITOR ANALYSIS

From this a much more detailed assessment of the other players in the field can be undertaken. This includes evaluating the initiatives that they take to promote their own marketing advantage and also to measure the likely responses to initiatives on the part of others in the sector and of those seeking to come in. The components of a competitor analysis are shown in Figure 3.4.

The outcome of this is a detailed discussion and analysis of the main competitors, at least in the field. Consideration is given to: strategy of the competitor; its driving and restraining forces; its current operations and capacities; its current marketing operations and activities; and its confidence and reputation. This is considered in relation to any general assumptions held about the competitor and about the industry itself. Detailed profiles of each competitor can then be drawn up. The outcome is further development of the strengths and opportunities present in the field and the ability to draw any lessons that are there to be learned from existing players in the sector.

This range of activities (or the equivalent) is essential as a prerequisite to establishing a competitive position. From this range of activities an indication of market strength and organisational capability is gained, and these can then be matched together. The next part of the process is to develop this into the establishment of a basis on which to compete.

ESTABLISHING A COMPETITIVE POSITION

Competition is the process of striving against others to win or achieve something. Competition always exists where there is choice. Choices are taken in varied and sophisticated ways according to means, circumstances and preference quite apart from any inherent or supposed nature and strength of the choices offered. The capability to compete is essential for the business of gaining customers and potential customers.

Competition has also to be seen as competing for scarce, limited and finite resources. It is affected by levels of disposable income and the extent to which the offerings are essential, desirable, non-essential, luxury, general, peripheral and marginal. It is a variable, extendable and continuous process, and subject both to universal outside pressures and also to local, regional and sectoral variations in the nature and levels of those pressures.

Profit

Profit is the consequence of conducting effective business. From the contractor or supplier's point of view, this means establishing volumes as well as quality of activities. From the client's point of view, this means using the different aspects of the industry so that a real or perceived profitable outcome is achieved. From the consultant's point of view, it means generating a sufficient volume of clients on the basis of the quality and integrity of the expertise offered, together with the ability to present this as a unique benefit to each individual client.

Profit therefore refers to the value and frequency of offerings sold. It is a function of income levels over periods of time. It is a function of the complexity, nature and sophistication of the products or offerings portfolio. It is measured in terms of rates of return achieved – across the sector; in absolute terms; by product; by product mix; in percentage terms; and in terms of sectoral norms and expectations; and in terms of timescale.

In terms of consultants and other experts and professional services, profit is the by-product of the capability to present the expertise and specialism in ways acceptable to and needed by the clients. It has reference to the nature and frequency with which the client base is serviced; the ability to command different levels of fees; and the wider perception of value with which such advice and expertise is held.

The competitive environment

The competitive environment is that in which competition takes place. It is affected by:

- The degree of captivity or the extent to which the market is captive because consumers have to buy from particular suppliers; or the extent to which it is fluid (whereby anyone can supply the market, and any consumer or client can go to anyone for supplies).

Figure 3.5 The competitive position.

- Whether the market is expanding, static, stagnant or declining, and each of these elements may be measured in terms of sales in financial or volume terms. They may also be measured in terms of the numbers of customers and clients, and in terms of behavioural aspects such as reputation, perceptions of quality, general confidence and the perceived benefits. There may also be dysfunction between the requirements of clients and the nature of the sector – for example, in much of the Western world there is great demand for civil engineering products and infrastructure developments while, at the same time, many governments are cutting back on their commissioning of this work.
- Activity volumes, which include the levels of income necessary to sustain profitable business; the levels of income necessary to support this; the levels of income necessary to maintain the confidence of backers and stakeholders; and the levels of income necessary to sustain investment for the future.

It is also necessary to consider the size of the client groups requiring the products or activities; the priority of the work in question relative to other demands; the acceptability of the activities both for their duration and in the final use of the finished product or project; and the extent to which this may be varied or client behaviour changed.

Assessment of the competitive environment is the prerequisite to the effective definition of more precise market segments in which activities are to be conducted; and from this, particular customer analysis may be carried out (see Figure 3.5).

SEGMENTATION

A segment (also sometimes called a niche, though the two terms are not precisely interchangeable) is part of the business sphere, environment, community, field of activity or operations that is both large enough, accessible and distinctive enough for an organisation to consider for the purpose of engaging in a range of profitable, successful and effective activities. The purpose of defining segments and niches is to understand the nature of that particular part of the market that is to be served and to understand the wants and needs of those within it so that the capabilities of the organisation can be presented in ways best suited to take advantage of these.

Segmentation is based on the following elements. The boundaries between each are not clear and sometimes more than one of these elements may be used to define a segment.

Social segmentation

In the UK social segmentation is as follows:

(a) aristocrats, upper middle class, higher managerial/directorial, administrative and professional;
(b) middle class, intermediate managerial and administrative, professional, technical; anyone holding official or professional positions in organisations;
(c1) lower management, clerical, white collar;
(c2) skilled manual workers such as electricians, building craft operatives, engineers; anyone with a trade or technical or technological training;
(d) semiskilled and unskilled manual workers;
(e) those without work related incomes: the unemployed and those on benefits and pensions of one sort or another; casual workers; the underclass.

This listing is based on the occupation of the head of the household (who was normally taken to be the male in the situation of a married couple or family). It is regarded as useful and convenient rather than an entirely valid basis for segmentation. It is not straightforward to assess segments on this basis alone (see Summary Box 3.2).

SUMMARY BOX 3.2 **Social Segmentation applied to the Construction Industry**

An equivalent form of this may be considered as follows:

(a) government and national public service infrastructure contracts;
(b) design and build, high and medium quality commercial facilities;
(c1) local government and health authority contracts;
(c2) regional road repairs, town centre refurbishment, commercial and domestic renovation activities;
(d) domestic refurbishment, e.g. double glazing, kitchens;
(e) essential works, e.g. storm damage, repairs, other activities in response to crises.

The purpose here is to relate the nature and volume of activities necessary in order to sustain long-term profitable activities. Towards the bottom end of the scale, it is likely to be necessary to be involved in more than one segment. Marketing has therefore to consider the differing needs and wants of those sectors rather than a simple presentation of general expertise. The converse of this is to be fully mobile – for example, in response to crises and damage if this is the only area of activities contemplated, the contractor has to be able to travel quickly to anywhere where such disasters do occur and very quickly to be able to make a distinctive and commercial presence felt.

Clearly, these distinctions cannot be all embracing. Divisions soon become blurred. However, it is a useful general marker for those considering marketing activities. It also gives a general indication of the likely patterns of consumption.

Demographic

Simple demographic measures of the population are age; gender; location; and density and distribution of the population. It is possible to segment markets on this basis alone – for example, it may be possible to be a long-term supplier of building materials and double glazing on the basis of being in a densely populated medium to high income area. It is also possible to sustain and grow a long-term reputation as a public sector contractor in a particular area because of the volumes of the population and on the basis that schools, hospitals, roads, factory and domestic infrastructure will therefore always be needed.

It is also possible to assess and decide what other needs they have. This enables the decision to be taken to make other offerings to the given sector. For example, those who work mainly on public sector infrastructure contracts may diversify into speculative house building or design and build activities for supermarket and industrial estate projects; or into public service construction consultancy; as this particular market changes (and above all, as it declines).

This may then be developed further. If the organisation succeeds to the extent at which it becomes a local brand – enjoying a high degree of local reputation – it may extend its range of services into domestic and commercial building products; design services (e.g. for domestic conservatories and extensions; public sector refurbishment) (see Summary Box 3.3).

SUMMARY BOX 3.3 **Population Sectors and Segments: Summary and Elements**

Whichever view of segmentation is taken, sectors consist of a combination of the following elements:

- **social:** groupings by income, age, gender, occupation and location;
- **generalised:** for example, house buyers, motorists;
- **specialised:** for example, executive home buyers, BMW users;
- **seasonal factors:** influences on buyer behaviour (for example, local roadwork commissioning in January and February; infrastructure development following a general election);
- **characteristics and appearance:** with reference especially to design, landscaping and planning;
- **fashionable, transient, faddish:** likely to require that facilities can easily be refurbished and upgraded;
- **national, international, cosmopolitan, global**;
- **supply driven:** for example, schools, public transport, hospitals, transport infrastructure;
- **demand driven:** for example, shopping centres;
- **crisis driven:** domestic and public building repairs;
- **technology driven:** materials production and usage; design and planning;
- **means of purchase:** purchase periods; credit arrangements;
- **means of payment:** preferred/desired/available.

These elements are common to all sectors and segments by whichever means they are defined at the outset.

Economic segmentation

This is the ability to group the population according to the amount of income of its members. This can then be developed so that patterns of spending can be assessed. The amounts of disposable income that client groups have are calculated, together with their propensity to spend; where they spend it, what they spend it on and when they spend it can then be measured. Major commitments that the client group has or is likely to have are assessed and deducted from the disposable income. Spending patterns can then be prioritised.

It should also be noted that economic segmentation varies extensively as the result of outside changes. For example, rises in interest rates mean that those with mortgages to pay have greatly reduced disposable income and this in turn leads to loss of sales markets and, in many cases, livelihoods in other sectors, as well as leading to stagnation in the property market. Reductions in public expenditure mean that those organisations dependent on public service commissions have to seek client bases elsewhere.

Buying patterns

Buying patterns reflect the nature, intensity and frequency with which purchases are made by clients. Intensity of buying relates to the volume and quality purchased at any one time. The frequency of buying is the number of times that this 'any one time' occurs.

The nature of buying refers to the fact that, for many activities in the construction industry, buyers have no particular expertise or detailed knowledge of what they are purchasing. House buyers want somewhere where they can live and enjoy a high general quality of life. Public works are commissioned by politicians, civil servants, local government officials and other public sector institutions to improve services and infrastructure.

Quality

This is the identification of the extent to which quality is a concern. In some sectors, this is a clear economic advantage for those able to satisfy high levels of demand for quality, which occur where customers are prepared to pay almost any price provided that the quality is supreme.

Elsewhere – again, especially in public service contracting – the ability to pay high prices is not universal. In the UK, there has been an overwhelming drive for public service work to be conducted to price rather than quality advantage; and this still remains the case in many instances. There are however signs of change in this – especially under private finance initiative and other equivalent arrangements whereby the finished facility is leased from the contractor by the particular public institution in return for very specific quality standards. Many institutions of government are now publishing their own highly specialised quality guidelines to which all potential contractors are expected to accede.

Whatever the basis on which segmentation is carried out, the purpose is to arrive at a much more detailed understanding of the particular area in which activities are contemplated. Attention can then be concentrated on specific customers and clients.

CUSTOMER AND CLIENT ANALYSIS

The basis of effective customer and client analysis is identifying the following elements:

- **Characteristics:** personal and professional characteristics, their general level of expectation; those features (in all commercial activities) that cause confidence in contracting and supplying organisations to rise and to fall.
- **Priorities:** their real priorities; their own perceptions of their priorities; the things that cause these priorities to change.
- **Expectations:** their expectations; their perceived expectations; the level and nature of activities necessary to meet and exceed those expectations.
- **Disposable income:** for major contracts and public service activities the equivalent of this is assessing and determining the budget that is available for activities; this has then to be related to the factors within and outside the control of the client or customer; and related also to the propensity to spend – for example, politicians mindful of debates on tax levels may continually reorder their spending priorities so that particular contracts never actually come up for tender and fruition.
- **Answering the questions:** 'why use you?' and 'why use your competitors?' This is a redefinition of the contracting organisation's own view of its competitive strengths. These must be checked with the customer and client groups. For example, there is no point in being the price leader in the sector if the client group wishes to enhance the quality or product extension, or high levels of service – for which they are prepared to pay precisely according to the nature and levels to which their own needs are satisfied.
- **Confidence and security:** the customer and client perception that once work is commissioned it can safely be left in the contractor's hands.

As the result of this range of activities, a complete understanding of the potential range of business activities, volumes of activities and income, and those elements that are going to cause clients and customers to choose you rather than someone else is now possible. Assuming that this assessment indicates that profitable and mutually successful activities can be generated, the next point is to translate this into a form of strategy that gives a sustainable competitive advantage (see Summary Box 3.4).

SUMMARY BOX 3.4 **Competitive Advantage: Examples**

Jobbing builders in a small town

There are three jobbing builders dominating the market in a small town:

- one is the clear cost/price leader, but there is normally a 3 to 4 week delay before he can do the work;
- one is the most expensive but has the reputation for top quality work and always clears up the mess he makes;
- one advertises only in the local Church of England parish magazine and 'as he advertises through the Church, he must be good'.

House builders in the same town

The two major house builders in the locality are as follows:

- one builds houses to a good standard of quality and also provides designed gardens, carpets, fitted furniture and kitchen, mortgage finance plan and guaranteed buy-back for the new occupier's previous house;
- the other simply builds the house to the same quality as the first and also includes double glazing as standard; there are no other extras; on average, prices are 30% lower.

Architects

Of three architectural firms in a semirural area:

- one has worked entirely on community activities and, in the past five years, has successfully designed out-of-town superstores on behalf of major national supermarket chains; local attitudes to these are negative though the stores themselves are very popular;
- one has worked on designs for renovating two local town centres that have lost prosperity as the result of the opening of the superstores indicated above; attitudes to these new town centres are very positive, but people still use the superstores;
- one works entirely for the local authority and health authority, designing new buildings and also producing plans for refurbishment of existing facilities; this practice enjoyed the complete support of the local political establishment for 35 years, though this is now uncertain because there has just been a change of political leadership and control in the area.

In each case, the strengths and weaknesses, the specific and distinctive advantages are indicated or inferred. This is also not an end in itself and will remain constant only:

- as long as customers and clients continue to be satisfied;
- as long as the existing competitors make no incursion on to each other's patches;
- as long as one or more do not go out of business;
- as long as there is no incursion from outside the area by new players;
- and as long as the clients continue – from choice or convenience – to offer them work.

Each therefore has distinctive areas of marketing activity on which to concentrate – and this concentration has the twin purpose of protecting the existing position and of looking for other opportunities.

Table 3.2 *Establishing a marketing strategy.*

Issue	*Product led*	*Market led*
Definition of the market	Markets are arenas of competition where corporate resources can be profitably employed	Markets are shifting patterns of customer requirements and needs which can be served in many ways
Orientation to market environment	The strengths and weaknesses relative to competition Cost position Ability to transfer experience Market coverage	Customer perceptions of competitive alternatives Match of product features and customer needs Positioning
Identification of market segments	Looks for cost advantages	Emphasises similarity of buyer responses to market efforts
Identification of market niches	Exploits new technologies, cost advantages and competitors' weaknesses	Finds unsatisfied needs, unresolved problems, changes in customer requirements, changes in organisational capabilities and expertise
Timescales	Fixed, stable, enduring	Fluid, flexible
Behavioural aspects	Reputation, image, confidence	Empathy, satisfaction
Competitive advantage	Expertise, quality	Flexibility, responsiveness

ESTABLISHING A MARKETING STRATEGY

The next step is to establish the best standpoint from which the market is to be approached. In order to do this, a view of the market must be established and this is illustrated in Table 3.2.

Marketing efforts can then be directed either at the excellence of the products and services on offer or at gaps in the market and the enhanced satisfaction that is to accrue as the result of conducting business.

Marketing strategies

The distinctive strategic position can now be considered. The premise for this is that no organisation can serve an entire market, segment or sector for all time; that competition therefore exists; and that customers and clients have a choice – at the very least, the choice to refuse or reject. Organisations have therefore to decide which parts of the market they are to serve, and to establish a distinctive basis as to how this can, most profitably, be achieved.

Porter identifies three distinctive positions:

- **Cost leadership:** the gaining of advantage through being the most cost-effective operator in the sector. This enables, above all, the ability if necessary to compete on price that is available as the result. In general, this is likely to produce the capability to offer standard, adequate and medium quality products and services in markets where these are the key characteristics required. The extent of the firm's success is likely to depend on the levels and price that can be commanded. Premium price levels clearly lead to the prospect of high margins and high levels of financial success and performance.
- **Focus or specialisation:** the offering of distinctive and often narrow ranges of products in a particular niche, seeking to serve all of those clients or customers in the particular niche, and seeking as many possible outlets for the distinctive product and service. This requires a basic concentration of identifying, anticipating and meeting the needs of the sector and ensuring that this is accurately completed. Profitable focus relationships are then normally based on product quality and the certainty and continuity of the relationship.
- **Differentiation:** in which the basis of business success is founded on the emphasis on marketing and advertising, and image-building activities, the purpose of which is to set the product apart from others in the sector and maintain the ability to sell at a premium price. The outcome is a uniqueness or identity for the products that are widely valued by buyers other than price advantage. In many cases, this involves the creation and maintenance of branding – the generation of a distinctive identity for a product, service or organisation name (this is discussed more fully in Chapter 5). Effective differentiation attracts the attention of potential customers and clients in response to their wants and needs – whether real, perceived or stated. If a customer needs quality then this becomes the selling point of the product. If the customer needs association with some preferred form of image – for example, identification with the company that designed the Lloyds Insurance building, Waterloo International railway station, etc. – then this becomes the point of view from which presentation is undertaken. If the customers need convenience, speed of delivery, ready access or easy payment terms, then in each case they will be satisfied only if these are produced as the key points of differentiation. This applies to some extent in building products and services, speculative house building, the financing of house purchases by building companies, and domestic and commercial refurbishment such as double glazing and facilities upgrading.

The hypothesis is that effective organisations and marketing strategies have to be clear about adopting one of these positions if they are to be truly effective. Only by doing this can further initiatives be successfully contemplated and undertaken. It is clear that each brings with it specific obligations. To be cost leader means continuous investment in state-of-the-art production technology, expertise and the ability to concentrate on those sectors where cost leadership can give maximum business advantage. In order to differentiate, extensive investment has to be undertaken in advertising, sales calling, promotional planning, access and convenience. Focus brings with it the obligation to continue to seek new outlets for existing products; and constantly to upgrade the range of products, services and expertise on which the focus strategy is based.

SUMMARY BOX 3.5 Preferred and Priority Contractors

Many client groups now have their own lists of preferred and priority contractors. These include public services, local and national government, theme and leisure parks, supermarket chains and other larger corporations and multinationals. The capability necessary to get on to these lists varies between sectors and organisations. From a strategy point of view, this may involve any of the three stated positions. To be a long-term public service contractor may require the ability to sustain long-term cost advantage in response to the ever-reducing prices and fees available in the sector. To be the preferred contractor for a supermarket chain or multinational organisation may mean the ability to present the perceived quality and advantage of doing business with the contractor as the client organisation continuously seeks quality improvements and enhancements. If an organisation gets a reputation for being only able to work in a very narrow field, such as the design of leisure and theme parks and golf courses, then it may have to extend its location of marketing activities extensively and become a focus operator in the field as other client bases perceive that it is no longer interested in other work. The key features within this context are:

- **Speed of response:** the clients want the work doing at their convenience and this normally means as soon as they have cleared the planning and preconstruction hurdles. The small jobs – such as a double glazing contract for a housing estate – the dates have to be worked out sufficiently far in advance to be able to notify the residents, though once this is done the contractor must stick to these dates.
- **Quality of response:** especially in the private and multinational sectors, clients expect the job to be 'right first time, every time'. Private sector clients also expect an assumption of responsibility for problems by the contractor and do not expect to get involved in lengthy wrangles that hold up the ability to commission and use the finished facility.

- **Range of response:** preferred contractors are normally expected to have capability (or instant access to capability) across the entire range of construction and building disciplines if this becomes necessary. For example, if a part of a facility needs redesigning, the preferred contractor will attend to this with a view to produce the new drawings for consultation and approval as part of the total agreement. Or if one source of raw materials dries up or a subcontractor fails, the preferred contractor is expected to have the capability to replace this without slippage in the job.
- **Costs and charges:** preferred contractors are expected to be able to give a final price for the job before it commences. This overwhelmingly applies when producing facilities for the private sector; increasingly, it also applies to public service contracts.
- **Speed of access:** preferred contractors are expected to be accessible to the clients at all times and this includes project planning, design, building, fitting out, after-care and maintenance. In turn, this means having a substantial knowledgeable and authoritative public relations and client liaison service.

In summary, it is essential, whichever strategic position is adopted, to be able to provide a full range of work and expertise to the required quality; and to generate a given quality of continued working relationship. At stake is a position of mutual benefit and profit. Capability and willingness have therefore both to be presented in terms that meet very high product, service and deliverability specifications.

Other views of marketing strategy have also to be taken.

Offensive and defensive strategies

These terms are used to describe the play/response activities undertaken by organisations operating in the same sector and their competitive and strategic relations with each other.

Offensive strategies are undertaken in the pursuit of extending market share or dominance at the expense of other players in the sector, or with the view to expanding the total potential of the sector. They may also be contemplated as one of the responses to be given to other offensive players in the sector: Player A seeks to gain market at the expense of Player B, so Player B attacks Player C and so on. The main features of offensive strategies are distinctive and aggressive marketing campaigns which normally have the purpose of 'trumping' the offerings of other players in the sector. Such strategies arise for a variety of reasons. They may be as the result of over-capacity in the market or in response to new entrants or potential new entrants and alternative offerings. They may also be the result of a more general sectoral lack or loss of stability. They may arise when a new player or potential entrant sees an opportunity (as for example, with a multinational civil engineering contractor entering into regional and smaller activities as the result of over-capacity). Still others arise where a radically different approach to the existing sector is contemplated – for example, the entry of commercially oriented design and build contractors as competitors in the public service and hospital building sector.

Offensive strategies may also arise as the result of technological improvement, volume and cost advantage, or from the control of supplies and raw materials, or outlets and the means of distribution.

Defensive strategies are undertaken in response to offences by other players in the sector. They involve one or more of the following. They may respond to the offensive activities of other players: for example, if the quality of one's operation is called into question, the defensive response is to rebuild any loss of reputation for quality. They may also concentrate on other distinctive elements of their own offering – for example, if cost/price leadership is no longer an advantage then the defence may be additionally shored up by attention to product or expertise quality. They may also concentrate on other features of the attacker; this means turning defence into attack – and results, for example, in price wars, dumping and flooding.

Organisations may ignore the offensive in so far as they continue to concentrate on their own strategy. If this is followed, it is essential that a watching brief at least is kept on the moves elsewhere in the sector. Price cutting may shift the whole range of margins hitherto available in the sector. Marketing activity may transform the image of the sector. Organisations must positively arrive at the judgment that they can maintain their position in spite of offensives.

It may also be that a particular marketing activity is seen as offensive by other players in the sector and that they make their own response. For example, if one civil engineering contractor decides unilaterally that it will extend its product into land and site refurbishment and the construction of related infrastructure, it may do this with its own purpose of transforming its own standards. However, competitors may respond by paying attention to the quality of the core product rather than following the contractor down the path of product and service enhancement.

Other factors have now to be considered.

Expansion and growth

Expansion and growth strategies take one or more of the following forms:

- Expansion and growth of market share – either related to expansion of the total market size or else at the expense of other operators.
- Expansion and growth of turnover, volume, income, profit and activities.

This may relate to market share also, or to a drive for efficiencies within the organisation, or in the seeking of greater levels and volumes of business within the existing customer and client base.

- The achievement of economies of scale and command of technology, distribution, supply and expertise.
- Command of the means of gaining and maintaining customer and consumer confidence.
- Expansion into new market areas and niches – where existing products and project ranges are priced or differentiated in ways designed to appeal to different niches. In highly competitive sectors, the requirement is to broaden the appeal of the offering in relation to those of other players. Related to this is often the diversification and extension of the offering range – either real or differentiated.
- Expansion into new locations – both indigenous and international.
- Diversification, integration, acquisition, merger, joint ventures and alliances. In international and some regional activities this includes the engagement of local partners.

With all expansion and growth activities the rational and the irrational must be considered. Rationally, organisations expand because they have spare or under-utilised resources, finance, staff, production and output technology; or because their current markets are no longer capable of sustaining their levels of activity. Some organisations may need to grow in order to remain viable players in a turbulent sector (as with national and global companies in the civil engineering sector). Organisations make acquisitions to seek more and alternative outlets for their products – for example, a civil engineering company acquiring a quarrying company. Organisations expand the production and output volumes by using their existing technology more effectively and thus bringing unit costs down.

From a marketing point of view, the irrational is much more critical. The expansion, growth or merger has to be marketed, both to the client base and also to the staff involved. Some organisations have overwhelming pioneering cultures that require constant prospecting and exploration. Organisations may choose to go down a particular line because it is a high profile or fashionable area in which to be involved, e.g. the UK civil engineering industry expanding its activities into the Far East. Organisations may choose to go into a sector because it is, or is perceived to be, highly profitable (rational enough) but without having any previous experience or expertise in it (again, this has been found to be a problem for civil engineering, architectural and building practices when they have gone overseas).

There is also the point that the marketing activities undertaken by an expanding company must be acceptable to the new sector or sectors in which it is to engage in activities. In the cases of takeovers, mergers, acquisitions, joint ventures and local partners, the marketing activities of each must be capable of being harmonised so that a mutually positive market as well as operational relationship are achieved. There is a derived point here – the staff involved in such ventures must be confident of its position in it, otherwise it will dilute any marketing initiatives proposed.

SUMMARY BOX 3.6 The Use of Buying Criteria as the Basis of Strategic Design

Success in securing business is reliant on meeting and matching the prioritised buying criteria of the potential client. This revolves around three key areas of:

Individual
Company
Product

Where the purchase involves securing a high value and high risk commodity, product or service, then invariably the client buys the individual, then the company, and finally the product.

These buying criteria are as a result of the following factors:

1. The type of sale relies on an on-going relationship between supplier and buyer, but more pertinently, between salesperson and purchaser within the client company. Therefore the need for social interaction is of paramount importance.
2. The nature of the purchase has a very 'public' implication for the purchaser in question. The failure to purchase the correct product will result in a public mistake. Therefore the impact of risk on the buying behaviour results in the purchasers focusing on elements on which they feel secure: these are primarily the individual they are dealing with and the status and presentation of the organisation that they represent.
3. The long-term implications of the purchase, and the initial capital-intense start-up phase required in some instances, mean that the purchasers must understand the value implications to justify their decision. This emanates from the individual's ability to analyse and develop the value justification 'within' the client's organisation.
4. Finally, the solution must be technically capable, but it is only at this stage that the product has any role to play in the decision makers' buying criteria.

These prioritised buying criteria dictate the order in which the customer views the sales process. Moreover, it is through each of these elements that the customer builds an opinion of the overall company that they are dealing with. Therefore, to create a truly customer-focused approach to business, every strategy in every area must assess the impact and role of each of the three principal areas. In doing so, the continuity of the value chain of the organisation is never compromised.

On this basis, each principal activity of the business, whether it is creating a new market offering, building market presence, recruiting and training or installing new processes, can be defined under the following framework:

1. **Individual**
   What are the requirements and implications on the individual?
2. **Organisation**
   What are the requirements and implications on the organisation?
3. **Product**
   How does it create and develop the product offering?

With a focus on addressing each of these areas, any strategic design is consistently referenced back to the customer, and the way in which the customer views the business.

Furthermore, by applying the idea of the 'internal customer', this analytical structure can be extended to influence every element of the organisation. With the internal interaction of each part of an organisation operating to the same buyer–seller standards promoted for the external operation of the company, an efficiency level is achieved through the consistent focus to performance and delivery.

Any element of an organisation's strategic design that falls outside this fails to concentrate on the customer as the focus of the business. Invariably, it is such elements that create needless bureaucracy.

Many organisations recognise the centrality of the customer in defining the business. There is also a recognition of the importance of at least one of the key buying criteria in building the appropriate analytical process in order to facilitate the process. Some organisations are able to build considerable competitive advantage using just this single element. Excellence in the individuals alone can secure the confidence of customers to purchase, even though delivery from the organisation is poor and the tangible product is only marginally different from that offered across the market place. In this instance, individuals fulfil the other criteria, through their representation of the organisation, and their technical knowledge and understanding, creating the 'value added' components of the product.

Since the 'individual' is the most significant element in the customers' buying criteria, this is an effective substitute for any shortfalls in 'company' and 'product' over the short term. Neither company nor product could have the effect since the importance of both is not as great. However, even the influence of the individual alone is relatively short lived. The failure of the 'organisation' and 'product' criteria to be fulfilled adequately creates long-term damage to reputation and builds a generally negative perception on the overall ability of the company.

Continually evaluating and re-evaluating an organisation's position relative to the three principal components of customer buying criteria builds a universally strong strategy. If customers view an organisation through particular categories, then this is the way in which organisations should view themselves. Only by looking at the organisation in the same way that a customer sees it, can true excellence be achieved. This allows an organisation to be built around customer criteria, and creates an uncompromised internal customer focus.

Source: A. Impey (1997) *Project Manager*, Touchbase UK Plc.

Market entry and penetration

This occurs in the following circumstances and conditions, and it should be noted that there are very few sectors that are not open to potential entry:

1. Where the overall market is growing, can be made to grow or has growth potential. In this case, there is a capacity or potential that existing players may not be able to fill. They also may not be willing to fill it, preferring to concentrate on keeping their existing client bases satisfied rather than diluting their efforts.
2. Where organisations leave the market, again leaving spare capacity.
3. Where the potential entrant has a real, perceived or differentiated advantage to bring to the sector in question. This is overwhelmingly likely to be a cost/price, value or quality advantage over existing operators.
4. Where a niche operator builds up sufficient reputation, resources and capacity to engage in offensive strategies with a view to taking a share from the main players involved.
5. Where complacency exists on the part of existing operators in the market; where there has been a fall off in the levels of quality or service; where profit taking exists; there are likely again to be opportunities concerning price or quality.
6. Where an organisation is able to bring its reputation from one sphere to another and to gain a foothold in the sector as the result. For example, the global perception of Japanese quality and value in cars and electronic goods has enabled building and civil engineering companies from that country to gain both favourable acceptance and also industrial footholds as the result.
7. The largest and dominant players in sectors are often able to increase their domination as the result of a combination of their commanding position and the consequent relative ease of building on their reputation. There are clear advantages to be gained from being in the position of dominant operator – especially those concerning the ability to penetrate those parts of the client base currently served by other players. If the dominant operator runs into difficulties, it is likely to have cost, quality or price advantages that it can present to the rest of its sector more easily than other players.

Consolidation

Consolidation occurs where organisations seek to preserve their positions in their markets, niches and sectors, range of activities and operations and client bases. Consolidation implies the reinforcement of existing positions. This is likely to be sought through increased marketing and public relations activities, together with improvements in efficiency, productivity and the cost basis of the organisation, and improvements in quality and delivery targets.

This may also be a prelude to seeking to maximise returns on investment in a market that is perceived to be on the point of declining. As the fact of the decline becomes apparent, organisations seek to gain every last possible benefit through 'strategies of harvesting'. This approach includes: sales of brands; sales of licences; sales of franchises; sales of technology; sales of primary sources and distribution means; client expertise and premises leasing; and the maximisation of main product/project sales. The purpose is to be left with as few unrealised assets as possible before the sector goes into decline.

Withdrawal, retrenchment and contraction

Withdrawal, retrenchment and contraction take place as the result of decisions to move out of particular activities and/or to concentrate the resources of the organisation on a reduced portfolio.

There is not necessarily a negative connotation to this. Withdrawal from particular spheres may take place as the result of reviews and activities that arrive at the conclusion that loss making or resource consuming offerings are being carried by the organisation's profitable and positive activities. It may also be a response to economic conditions: for example, where charges and interest rates are sufficiently high to make it worth subcontracting specialist activities rather than owning them.

Organisations wishing to divest themselves of loss-making activities must first consider the effect on their remaining range and on their continuing image, prestige and confidence. Concentration of resources on a more limited range of activities may lead in time to opportunities for expansion and intervention elsewhere. Within this context, withdrawal takes one or more of the following forms:

1. Concentration on core business at the expense of niche or peripheral operations. Much depends on the criteria used to define 'core' – and this may relate to volume, profit margins, income or resource consumption; and also to reputation, confidence, image and identity.
2. The end of the useful life of a hitherto profitable product, project or range of activities. This may come about as the result of the existing business having been superseded by a superior, more modern or fashionable alternative; or as the result of technological development that makes the current way of activities unviable; or as the result of change

of expectations – where one player has been able to reduce timescales for delivery or project duration substantially, other players (especially those not in a dominant position) may have to follow or else come out.
3. Where the organisation is faced with massive investment, transformation, modernisation or rehabilitation bills on which it cannot see an adequate rate of return.
4. Where activities take place in volatile areas, regions and sectors. For example, organisations operating in the Middle East and Central Africa know that in extreme circumstances that may lose their entire capital investment if it is nationalised or destroyed by the government of the country in question or if there is revolution or war.

Diversification

Diversification may be classified as either related or unrelated. It occurs again when either the organisation has spare capacity for which it seeks further outlets; or because operating in one sector gives rise to additional opportunities within it.

Related diversification

This constitutes the development of the range of offerings and activities within the current ambit or experience of the organisation according to its current expertise. It may be a direct relationship as, for example, where an industrial building company seeks to diversify into the house building sector. It may be a generic relationship, as, for example, where a domestic house building company seeks entry into the protected or sheltered housing sectors. It may be related to the capacities of the technology and technical expertise: for example, the offering of in-house consultancy services on a broader commercial basis. It may be an extension of the use of production technology owned by the organisation: for example, components produced overtly for the building and civil engineering industries are produced by technology that can easily be modified or rejigged to produce equivalent products for other sectors.

Unrelated diversification

This is where the organisation treads its own new ground. In civil engineering and construction a version of this is moving from the domestic market to become an overseas player. In such cases where the expertise remains the same, its application is certain to be very different. Mining and quarrying companies have diversified into the domestic furnishing and jewellery sectors – for example, the production of slate ornaments or wheal-jane jewellery.

Collaboration

Collaboration takes three main forms: joint ventures and consortia; hook-ups, associations and networks; and local partners. These occur where two or more organisations pool their resources for a project, piece of research or initiative.

The overriding concern is the mutual commercial benefit that accrues: the different organisations concerned pool their resources, expertise and technology in order to carry out the work in hand to the highest standards of each of the collaborating organisations. They are found in all sectors of the construction, building and civil engineering industry. The approach also enables elements of risk to be shared out between the member companies. It is also perceived by organisations going into a new area for the first time to be a relatively safe way of gaining 'toe in the water' experience of areas in which they have no previous knowledge.

CONCLUSION

The purpose of taking this approach is to build and strengthen the organisation as a going concern. From this form of assessment and in devising 'a strategic standpoint' the result from a marketing point of view is to combine what the organisation is good at, together with how it can be best presented to the client bases in question.

Implementation

Critical success and failure factors

Critical success and failure factors have often to be tested in some form or another. The usual approach to this is:

- Feasibility and pilot studies, often in the form of a mini-launch of a particular product, project or initiative. This is often accompanied by professional and national media coverage. These occur with pre-set objectives and criteria in mind, and also with a wider general assessment brief.
- Performance forecasting and projections, often based on computer modelling and simulations. These will be subject to intense scrutiny

and enquiry and must always be supported by assessing the range of best, medium and worst outcomes.

- Projections of total success should be considered: in extreme cases, for example, this could lead to the inability of clients to gain access to the particular product, project or expertise, and ultimately to loss of reputation.
- Projections of total failure – the nightmare scenario – should also be assessed so that the organisation is under no illusion as to what the total level of loss could be.
- Ultimate consideration – every product, project and initiative have elements that, if they cannot be overcome, mean that they are certain to fail. Whatever these are, must be identified, isolated and tested and ways found around them if success is to be achieved.
- Profitability and effectiveness assessment which is carried out over the anticipated lifetime of a product or project, and also identifying key and critical factors, elements and timescale punctuation marks along the way.
- Acceptability – the extent to which the particular initiative meets client requirements of fulfilling expectations, giving satisfaction and value as well as meeting price, quality and value expectations. In some cases, ethical and social factors are a part of this.

Beyond these, a general organisational strength and capability must match the demands of the direction proposed and be able to reconcile and accommodate these alongside the general range of activities and obligations.

More direct financial measures may also be required or demanded by the organisation as a condition for engaging in particular initiatives.

The most common of these are:

- **Return on investment**: the rewards gained as the result of placing resources and energy into particular areas and activities.
- **Return on capital employed**: similar to return on investment, normally measured against the net share capital of the organisation.

Marketing strategies

Once the general strategic position has been assessed, marketing strategies can then be designed to assess the following:

- Marketing mixes of all the products, projects and expertise to be offered. This is the combination of price; promotional, presentational and advertising aspects; direct sales activities; presentation of benefits; and the nature of locations and outlets where it may be obtained.
- Development of the organisation as being safe and steady and the promotion and presentation of confidence and strength.

- Determination of the sectors in which activities are to be conducted so that satisfaction, security and confidence may be presented in ways acceptable and positive to those client bases.
- The creation of activities in support of this including advertising campaigns; sales teams; brochures; information and public relations; help and support activities; and general activities in support of generating positive images and responses.
- Joint activities, where appropriate, so that whatever marketing activities are undertaken are capable of harmonisation by all those involved in the joint activities and so that the joint approach is acceptable to the client base.
- Market research and investigation to ensure that products, projects and activities are going to be effective and will remain so.
- Product life cycle assessment and the implications that arise – both in terms of where income, cashflow and profits are to come from in the immediate future; in the long-term future; and as a prelude to developing and improving both the existing range and also in seeking new products, projects and activities.
- After-sales support and services, so that whatever is done continues to give good and lasting value. As stated elsewhere, there are clear implications for price – whether a total price is to be charged or total project or service, or whether there are additional price considerations for each component part.

Acceptability

The acceptability to all those involved must be assessed and this includes all the organisation's stakeholders. Especially customers, consumers and clients must be satisfied that they are to receive continued high levels of service and quality whatever is undertaken. If an organisation expands, its clients need to be certain that they are not going to lose any feeling of personal relationship now that the establishment is larger – or that the relationship will be developed to be even better. If the organisation contracts, they must be satisfied again, that this is not going to affect either capability or confidence.

This is the background for all successful marketing activities. In simple terms, by giving a purpose and clarity of direction to the organisation, marketing activities can then be designed in support of this. If such purpose and clarity are not present then it becomes very difficult to devise and design marketing activities – or indeed anything else – and to have any clear criteria against which to measure success and failure.

And, again, this applies across all sectors of the industry. Also, while jobbing builders and other small companies may not think in precisely these terms, those that are successful have a very clear view of their own strengths and how these are best presented to the customers and clients that they serve. For larger companies and global players, time, trouble and resources are invested in assessing organisational strategy much more formally so that when marketing activities are devised, they are in accordance with the overall drives of large and diverse organisations.

4 Price

Agreeing the price for a piece of work seals the relationship between contractor and client. It is the central feature, and reflects the value placed on the relationship. It indicates strongly the quality, volume and duration of work required and the amount of commitment entered into by each party. It may also indicate a propensity to re-engage the relationship for subsequent activities and projects, provided all goes well.

Price is based on a combination of:

- what the market expects;
- what the client can afford;
- the nature of competition in the sector;
- what the contractor can afford to work for;
- any specific factors concerning the particular job, e.g. time, quality constraints, location, materials to be used;
- what is being agreed and paid for – whether the completion of a building or facility; or a wider process often supported by maintenance, presales and after-care agreements.

Different approaches to pricing are as follows:

- **Price tagging**: often called the retail view – this is where products and activities are given a stated price and the organisation is fixed to this. For retail and domestic sales of building products this is sufficient. Some civil engineering contractors also use this approach (e.g. to design, construct and deliver a road bridge 30 metres long and four lanes wide costs £x); and client bases in some parts of the world are happy to work to this. It is however much rarer than it used to be. Specialist agencies, consultancies and subcontractors also use a set price per hour (or per person-hour) for their services in many instances.
- **Indicative pricing**: in which a broader view is given of the going rate for work which, nevertheless, leaves open the need to negotiate precise agreement between contractors and clients for specific pieces of work.
- **Contract pricing**: in which agreements are made for specific pieces of work based entirely on their own merits and which represents the meeting point between the anticipated value to the client of the completed facility, the price that the client is prepared to pay in order to accrue that value and the level of charge for which the contractor is willing to do the work.

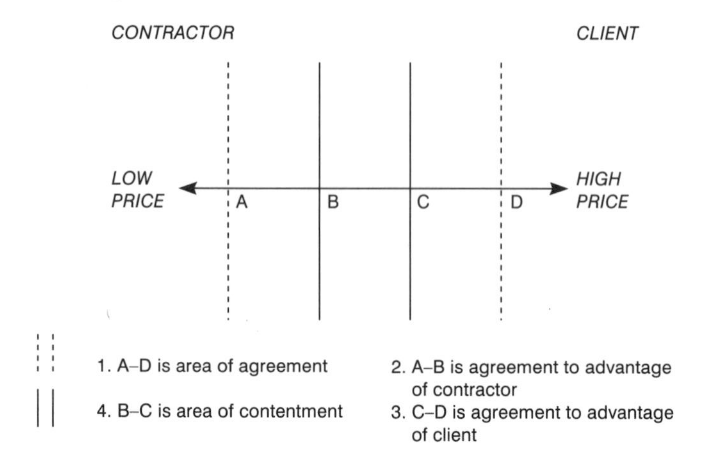

Figure 4.1 The negotiated view of pricing.

The relationship may also be seen as follows:

- **Consultative**: in which a joint agreement is reached openly and which relates to the real needs and concerns of each party.
- **The integrative or collaborative**: in which the contractor and client work out the project details and specifications jointly and then agree the price as the result.
- **The collective bargain**: in which the price is extensively negotiated, normally from a point of view of mutual mistrust (see Figure 4.1), so that what is eventually agreed is more or less acceptable to all concerned. Negotiation is based on fundamental mistrust – the opening offer/change is always: (a) so high (contractor) that the client rejects it; (b) so low (client) that the contractor rejects it. The aim is to get first to the range A–D; long-term relationships are based on B–C.
- **The dominant partner arrangement**: in which either the contractor secures a very high price from the client in the knowledge that the client has no choice (for whatever reason); or the client secures a very low price from the contractor in the knowledge that the contractor needs the work. From either point of view, this relationship is not normally sustainable in the long-term, though it is often used for known one-off contracts.
- **Itemised pricing**: where each and every item included in the contract is accorded a given level of charge – often this goes alongside the client seeking alternative contractors for different parts of the work.
- **Process pricing**: where overall responsibility is given to the contractor for the whole construction process, which may include environmental restoration and infrastructure connections at the completion

of the project; and which may include maintenance, refurbishment, extension, transformation and other post-completion activities.

- **Contractor–client joint ventures**: in which the project is jointly underwritten during its completion after which the rewards are then also jointly shared (or according to the proportion of risk undertaken by each party).
- **The private finance initiative approach**: in which the client (invariably a government department or agency) gives permission for the facility; the contractor then builds it and fits it out, underwriting it himself; and then recoups this investment through charging the end-users.

COSTS

Whichever approach is taken to price, the overall aim is long-term security based on a surplus of income over expenditure. It is essential to know what costs have therefore to be covered and how these are best broken down. From both a marketing and managerial point of view, the best approach is as follows:

- **Fixed costs**: these are incurred whether or not any work is done. Fixed costs include: staff; equipment and capital goods; property and accommodation; support functions; and marketing.
- **Variable costs**: these are incurred once work is commenced. Variable costs include: equipment usage and maintenance; materials and supplies purchase; energy; staff overtime and staff expenses (but not salary – this is a fixed cost); acquisition of staff equipment and technology for short-term purposes and specific activities.

The general rule is that so long as prices are set to more than cover variable costs, a contribution to fixed costs is made and that the greater the ability to charge above variable costs and for the greater the volume of work carried out, the fixed costs are covered and genuine profitability is achieved.

It should also be apparent from this that fixed costs represent a continuing and long-term obligation that, in turn, provides the basis for genuine commercial activities. A fixed cost base can be varied, long-term, should always be as efficient and effective as possible, and should indicate the level and quality of activities necessary to sustain the organisation into the future. Beyond this, the greatest mistake that organisations make is to see costs and profits purely in terms of annual and other accounting cycles. Annual reports and accounts are produced to meet specific legal and conventional requirements. Marketing is a continuous long-term investment and does not come to fruition or failure on specific dates or over periods of time that are regulated by the finance industry for its own reasons.

PRICING CONSIDERATIONS

For the construction industry, there are many considerations:

- Capital projects and building activities are priced so that variable costs – the costs incurred as the result of gaining the work – are covered. Ideally, the price also makes a contribution to the fixed cost/total cost of the contracted organisation. It is also necessary to consider charges incurred by the contractor as the result of having to underwrite the project from inception to delivery and from the need to acquire any specialist equipment and expertise necessary.
- Building products are priced so that an individual perception of quality, value and service are given and at a level that the size and nature of the market can sustain. It is also increasingly common to find differentiated pricing approaches (see below) as competitors seek real and perceived price advantages. This has especially been the case in recent marketing of domestic building products in the UK where, because of the depressed housing market, customers have concentrated on improving and upgrading their existing homes rather than moving. This also applies, to an increasing extent, to the refurbishment and upgrading of commercial and public service facilities.
- Building services, consultancy and agency provisions are priced according to expectations and perceptions of the nature and quality of expertise demanded by the client base and which the specialist, operator or consultant wishes to present. It is necessary to understand the client point of view here – there is a very distinctive balance of price, quality, value, volume and time as well as distinctive projected results that is being paid for. Especially where charges are high, clients expect their problems to be solved as the result of using a particular service. Charges also reflect the quality, scarcity and demand for the service on offer.
- Building materials are priced according to quality, volume and delivery methods; and the latter has specific regard to frequency and regularity. Depending on the nature of the work, a price may be agreed with a quantity surveying agency or directly between the quantity surveyor or purchasing officer and supplier.
- Architecture and design services are priced much more specifically. There is likely to be a fee for an initial proposal; a fee for a revised or final proposal; and a fee for the proposal once accepted. There is again a price, quality, value, volume, time mix; and, again, depending

on the standing and reputation of the practice, charges reflect the perceived and actual quality, scarcity and demand for its services.

The wider state of the industry and the pressures on those who commission work have also to be considered. In all construction disciplines, prices rise when there is too much work chasing too few companies; and the converse – where there is too little work chasing too many providers – drives prices down. The state of the industry in different parts of the world and different sectors and segments within those parts clearly varies, and that is why the ability to charge an effective and sustainable price level is dependent on full current knowledge and understanding of the markets in which activities are to take place.

Client pressures on price

The value chain

The value chain breaks an organisation down into its component parts. This is in order to understand the source of, and behaviour of, costs and actual and potential sources of differentiation. It also isolates and identifies the building blocks by which an organisation creates the offering of value to customers and clients. The concept was first devised by Michael Porter as part of the process of analysing organisations' competitive advantages. The hypothesis is that an understanding of strategic and marketing strengths and capabilities must start with identifying where value is gained and lost. These are identified as follows:

- **Inbound logistics**: activities concerned with receiving, storing and distributing components, supplies and materials that are to be used during productive activities. In construction marketing terms, this means attention to the quality and accessibility of raw materials – both those used in the direct pursuit of the client's interests and also those used during internal administrative and support functions.
- **Operations**: which transform these various inputs into the product, project or service in question. In construction marketing terms this includes the speed and quality of operations, the need to remedy faults, the availability of staff and technology, and the speeds and quality to which they work.
- **Outbound logistics**: the delivery of products, services and projects to customers and clients. Again, this includes time – from inception to delivery; and from order to completion. It also means recognising those things that can and cannot be controlled and the reasons for this.
- **Direct marketing and sales**: the means whereby customers and clients are made aware of products, services and projects and are able to gain access. This covers all direct marketing, distribution, public

relations and awareness raising activities – and, again, refers to the general marketing and public relations role of those not directly involved with customers and clients as well as those who are.

- **Service**: this covers all those activities that enhance/maintain the value of a product, project or service during the course of its useful life. This includes installation, repair, after-sales, preventive and crisis maintenance and, where necessary, the provision of spares and components. Increasingly, it also includes refurbishment and attention to alternative site usage. It includes the interaction between the finished project and its environment. Where appropriate, it also includes wider aspects of facilities management. For major projects built under private finance initiative and other contractor financing schemes, it means developing the expertise to deal with customers, consumers and other end-users.

As well as conducting this form of analysis within organisations, it is also possible to extrapolate the total process so that every point from inception to completion where value is gained or lost may be identified. Effectively, this means that value analysis is being carried out on all organisations concerned with the product or project in question.

Value management

Value management is a more precise and numeracy based approach to the costs, benefits (and therefore value) of specific products and projects. To be fully effective, it is therefore carried out by those with precise numeracy skills such as estimators and quantity surveyors. The purpose of value management is to *'maximise the functional value of a project by managing its evolution and development from concept to completion through the comparison and audit of all decisions against a value system determined by the client or customer'* (Kelly and Male, 1993). This means paying direct attention to costs, quantities and the quality of each element that is to be used in a specific project.

Clients are increasingly interested in this approach because of the great increase in the numbers of contractors from all disciplines available to them. Specific client concerns are as follows:

- the need to drive down costs;
- the escalation of estimated costs;
- where tenders are received in excess of budget;
- variations in the estimate of delays and logistic difficulties on the part of different tenderers;
- the relationship and co-ordination between designers, consultants, planners, co-ordinators and project managers;
- the need for independent audit or appraisal of the project; the desirability of independent audit or appraisal of the project;

- pressures on the client to achieve distinctive and/or prescribed profit margins;
- general needs or desires on the part of the client to seek alternative solutions and designs to projects;
- the role and influence of consultants in the client organisation.

Contractors and consultants have to understand the client's perspective on this. Additionally, they may also adopt value management techniques for their own specific reasons as follows:

- the drive continuously to improve all aspects of their provision – and this includes time, cost and quality;
- as part of the understanding of the changing needs, wants and demands of the client base;
- as part of a drive to improve the innovation and creativity of their own staff;
- as part of a drive for competitive advantage (possibly pre-eminence) through attention to value (and especially client value).

Value management activities are therefore normally carried out at the inception and precontract stage of projects. Increasingly it is essential that some form of study along these lines is carried out by all potential contractors and consultants who seek to be involved. The difference between this and the Porter approach is that it pays much more direct attention to costs and estimates; whereas the Porter approach tends to concentrate on the presentational, perceptual and service aspects of where value is gained and lost.

Price as a driving force

If a contract is to be awarded purely on price, it is as well that perspective tenderers and contractors know this in advance as they can then decide whether they are prepared to compete on what is likely to be a very tightly costed operation.

Price over contract duration

Ideally, also, contractors wish to know when they take on a contract that it is to be fully funded over the period to completion. In practice this is not always possible (e.g. the construction of the Channel Tunnel and the consequent pressures on Eurotunnel and TML). It may also happen that for long-term publicly funded projects and activities, funding is cut unilaterally part of the way through (e.g. long-term motorway repair and upgrade contracts). Or it may be that, in practice, funding is not available to do the job in an ideal way so a diluted or substandard version of the project is commissioned, or else it is completed piecemeal rather than as a whole (e.g. the M25 London Orbital motorway).

Production processes

Product in this context is where the client chooses (or is pressurised) to commission a product or project in isolation from its broader environment or usage when finished. When this happens, price pressures are limited to completion dates, quality of materials and site management.

Process here means that the broader view is taken – and this perhaps means including fitting out, environment restoration, after-sales and possibly also subsequent redesign, refurbishment and upgrade. There are heavier pressures on price here, not least of which is that because of the more complex and long-term contractor–client relationship, there may be radical changes in the cost for new fixtures and fittings and facility servicing charges.

Knowledge and influence

This is important where those who have the influence to commission work have no knowledge of the work itself. This applies especially to public works and projects that are overseen during work in progress by politicians and others who have responsibility for public bodies and institutions. It is at its worst where politicians and officials set a price for the project that is simply unworkable, which bears no relation to the ability to produce even a basic version of what is envisaged.

The converse is also true – where a contractor and client agree a price for the work between professional people, with full knowledge and understanding of the industry, but where the client's professional staff have no influence in the contract awarding and pricing process.

Time as a driving force

All contracts are time constrained. Time becomes a pressure on price where the job is extremely urgent, for whatever reason, and involves the contractor in additional variable costs, especially technology, plant, equipment, design and people. There may also be a pressure on the contractor's purchasing systems and policies; while the volumes required for the job are absolute, however quickly or slowly the work is carried out, the speed of activity required – especially when things are speeded up – is likely to create pressures on delivery schedules and storage space.

Expert and consultancy services, and architecture and design practices normally include speed of response in their pricing packages – though again, they may expect an additional premium if the demand is for instant or very short-term service.

There may also be a premium necessary where the contractor (from whatever part of the industry) is being asked to put a large proportion of its total resources available for the particular client for the duration of the activities.

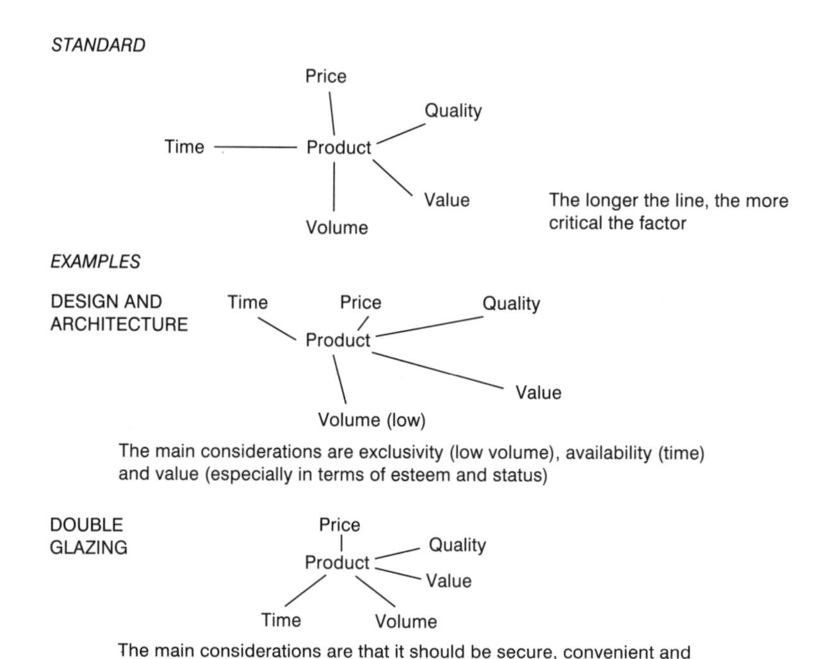

Figure 4.2 The price, quality, value mix.

Quality as a driving force

Where quality is the driving force, the price normally rises; so do expectations and satisfaction levels. The outcome ideally, therefore, is high-quality, high-value work for high prices and charges, and satisfaction. Satisfaction is not achieved or is diluted where quality and value do not match up with the price.

Figure 4.2 shows the mix of price, quality and value.

Perceived value

Given other constraints, the client engages the contractor with the view to achieving mutually successful high-value work. The contractor's name is associated with the project for the duration of construction and may last beyond its completion. The client's name is associated with the project long into the future and may be involved with it during completion. The price consideration here is therefore to understand that enough money is to be available in an ideal world to ensure that the result is indeed positive to all concerned; and to recognise and understand the consequences of being associated with something that is specifically constrained in this way.

Moreover, the perceived value may change and this may make the proposal very expensive in terms both of price and also continuing reputation. This happens, for example, when a contract that is attractive and agreeable to both contractor and client runs into difficulties either with the public at large or with the media (or both). It also happens if, and when, it becomes apparent to the client that the facility is no longer publicly acceptable or a priority during the period of construction, or if it does not fully realise its potential in the subsequent period of usage. In either case, there is likely to be an adverse affect on any continuing relationship – whether justified or not.

Pricing policies

Life is most straightforward when the pricing policies – pressures, structures, specific charges and overall attitudes to price – are known and understood by both contractor and client in advance of any work being commissioned. Decisions on price, acceptable levels and price coverage are agreed much more easily on the basis of full knowledge and understanding. If there are specific constraints – especially from the client's point of view – then the contractor can either write these into the proposals or else present its own version of the pricing approach in the expectation that other expertise will cause the work to be commissioned anyway, or choose to write itself out of the bidding.

Pricing principles

The principles of pricing are the same for the construction industry, therefore, as they are for all sectors. Specific applications vary however. The main principles are as follows.

Variable cost coverage

Pricing normally at least covers variable costs – the costs incurred specifically as the result of doing the work or carrying out the given project. Anything that can be charged above this makes a *contribution* to the organisation's fixed costs.

Total cost coverage

Some pricing policies arise as the result of apportioning a part of the fixed cost to each activity. The advantage of this is that the contribution of each activity to fixed cost, and therefore total cost, is known in advance.

The disadvantage is that it sets higher initial prices and this may have to be supported by some apparent or perceived advantage that will ensure that the higher charge is sustainable.

Penetration pricing

This is where the price is set low enough to guarantee a sufficient volume of work for the future from the given market or segment. The expectation is that once entry has been gained, prices can be raised to ensure profitable activity over the long-term. This is accompanied by guaranteed volumes and quality of production or service, presented in such a way that the market will buy from the new entrant rather than from existing players.

This is a long-term policy based on extensive investment and research if it is to be successful. It is also essential that prices can be raised at some time in the future in order to recoup the outlay; or that once one product has been used to establish a foothold, others can then be sold to the new customer base at a genuine commercial level (see Summary Box 4.1).

SUMMARY BOX 4.1 **Penetration Pricing**

Dumping

Dumping is the term given to the practice of flooding markets with products at very low prices with the express view of driving other competitors out of sufficient amounts and volumes of business for the 'dumper' to gain a foothold, to establish and develop a reputation; and this is also often alongside the business purpose of the 'dumper' to become the long-term dominant player in the sector. Dumping means that the prices charged for the products are not sustainable by the other players in the sector – and these either have to grit their teeth and bring their own prices down in the hope of driving the new entrant back out of the sector; or else they too have to seek other outlets.

Buying work

This is a less extreme form of dumping and it occurs when a contracting organisation can see a long-term and viable future in a particular sector but only if it can first of all gain a foothold and price advantage. For the immediate to medium term, therefore, it prices its work at a loss. The effects of this are not so extreme as dumping – when some of the players will normally be driven out. It is sufficient: (a) to make the new entrant sufficiently attractive to the existing client base; and (b) to cause existing players to consider their own margins again.

In other circumstances, contractors work at a loss because they understand or perceive that the long-term prognosis for the market or segment is good if they can only stay in business in the short-term; they are effectively taking the view that it is better business to work at a loss – and therefore to buy work – rather than to withdraw altogether.

Luxury pricing and economic rent

This occurs where:

- a contractor/supplying organisation perceives itself to be sufficiently expert and held in sufficiently high esteem in the industry to set a level of charges far in excess of other players;
- the client base perceives that, by paying the high level of charges, it is to receive the quality, value, volume of product, project work or service that makes it well worth paying the extra.

Economic rent is normally paid by clients to top quality and/or high prestige architects, designers, planning and contract consultancies. It is also paid increasingly to builders and civil engineers in turn for whole process contracts (see Chapter 7) where the client has sufficient flexibility to do this.

Economic rent is also normally paid by clients when they want work carrying out and completing quickly to the same levels of quality and value. There is an expectation of a price premium here – and indeed not to charge such a price premium in return for working under this form of pressure may cause a reduction in perception of value and excellence on the part of the client base (see Summary Box 4.2).

SUMMARY BOX 4.2 Luxury Pricing and Economic Rent

Pricing and branding

There are two ways to build a business: one is by offering low prices; the other is by charging high prices. Many managers, and all governments, think that low prices are the key and focus on keeping costs down on flexible wages and rationalisation. But a recent study listing the world's twenty most valuable brands identified the one factor common to them all – their continued ability to charge and receive high prices in return for the products and services.

Successful companies charge high prices based on high and continuing long-term levels of investment. Building successful brands requires a four-stage investment strategy. This starts with making a quality product but this alone is not the deciding factor. True successes go further – built around their products are clear brand features that differentiate them and add value. These include design names, symbols, reputation and marketing messages which create emotional associations as well as demonstrating the physical properties of the product.

The third stage is building the augmented brand. All successful brands add to their products personal services, loyalty schemes, guarantees, after-sales service and financial terms that provide additional benefits and support to customers and clients. With basic products being increasingly similar in quality, it is in this area of augmentation that much of today's competition is fought. The last stage is that when companies maintain this consistent and high level of investment over a long period, they achieve what is called 'brand potential' – in which a much higher level of the price/quality/value mix is achieved; and which engages much higher levels of customer/client loyalty than those achieved by cheaper competitors.

This approach to pricing is very little developed within any discipline of the construction industry. In very few sectors does any client base approach a company consistently over the long term on the basis of absolute confidence that, in return for the high price, an increased level of value and performance will be achieved. This is in direct contrast to the leaders of the consumer goods sectors – where, for example, Coca Cola, Levis jeans, Sony electrical goods all outsell cheaper supermarket or department store product equivalents.

The reason for this is that, with very few exceptions, no company in the construction industry has seen fit to invest in developing itself as a brand – brand building is seen as expenditure on advertising, design and customer support, as costs to be minimised rather than investments to be optimised. Also brand building does not sit well with the short-term focus that increasingly dominates British boardrooms – both in the construction industry and elsewhere. The emphasis is rather on cutting costs, which is perceived as a more certain and quicker route to short-term profit than investment in branding – which, especially for the construction industry, might not pay off for a decade.

By contrast, those that have built successful brands recognised a generation ago that the real route to long-term and sustainable success is through long-term investment which makes their products and services not cheaper, but more desirable.

Source: Peter Doyle – 'From The Top' (*The Guardian*, 21 December 1996)

Perceived value

Price has to reflect value. Different levels of price give different perceptions of the quality of product or service that is on offer. There is a universal perception that if the price is pitched too low, the product or service will be seen as cheap and therefore bad value rather than good value; and the particular company will do no business.

Perceived value is a reflection of the specific demands of particular market sectors and segments, and this value mix therefore varies (see Figure 4.3).

Price coverage and breakdown

This reflects the organisation policy standpoint of choosing whether to itemise and charge for each element of the contract or relationship; or whether to price and charge for the total service, process or package.

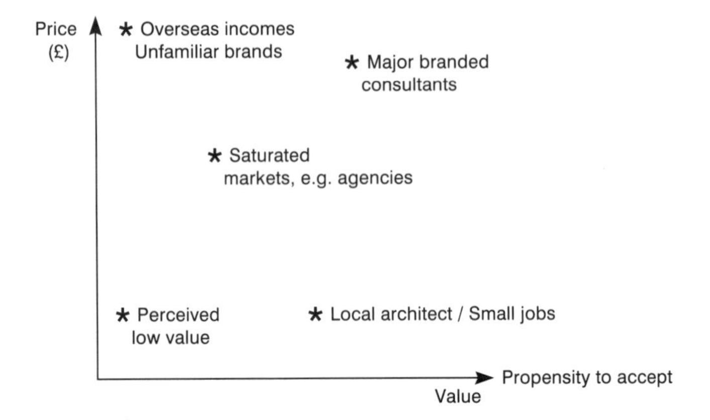

Figure 4.3 Perceived value and price example: consultant's charges.

Where there is any price sensitivity or flexibility, there is a move towards a total price for total service or process and this extends across the whole industry. For example, the purchase of building materials normally includes delivery; double glazing, kitchen and bathroom, and other domestic contracts increasingly include installation as well as the sale of units; while at the macro end, design and build contracts normally include the whole process from initial commission to commercialisation of the finished facility. The converse – of itemised pricing – is still extensively required and remains the norm in public service and government contracts (see Figure 4.4).

Other factors

Stakeholder pressures

Implications for pricing are as follows and have to be reconciled:

- **Staff**: expect levels of charges to reflect the drive for long-term, secure and profitable work that, in turn, secures their future employment prospects and economic well-being.
- **Shareholders**: expect annual reports and other published materials and commentaries to show levels of short-term profit and performance that enable dividends to be paid and increase the value of the shares.
- **Customers, consumers and clients**: expect to pay prices that are sufficiently low to give the perception of good value, and sufficiently high to secure the long-term future of the contracting organisation so that after-sales, maintenance and new product or project work

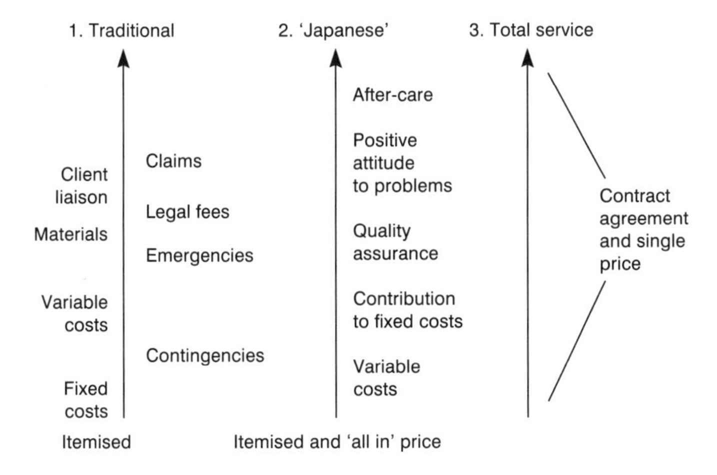

Figure 4.4 The pricing process: summary.

may be provided as necessary or desirable – and this applies to all construction activities.

- **The community**: communities consider this from three points of view –
  (a) for public service projects that they are receiving adequate and good-quality facilities and infrastructure in return for expenditure from taxation;
  (b) for commercially commissioned projects that they enhance the range of commercial, social and domestic choice and general quality of life;
  (c) that all activities, macro or micro, public or private, enhance the general prosperity of the community.

  Communities also generally expect that all construction activity will enhance the quality of their environment, the quality of working, social and domestic life, and provide them with an area of which they can be proud. These are indirect pressures on price but nevertheless reflect both general and specific expectations.
- **Client pressures**: the two extremes of this are – that the client has absolutely no leeway on price and the job is therefore done to whatever quality the price level can achieve; while at the other extreme, the client will pay whatever price is necessary to ensure that the desired quality of finished product or facility is obtained (see Figure 4.5).

Between the two extremes, there are some useful indicators:

- if the price is too low, nobody of quality will be attracted to do the work;
- if the price is too low, it is impossible to expect high-quality work in return;

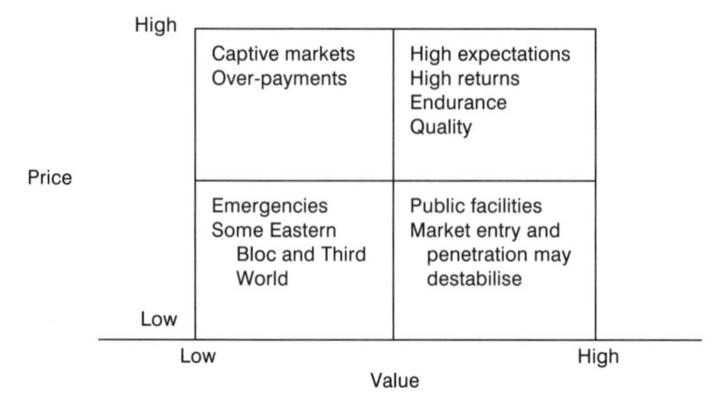

Figure 4.5 Client pressures.

- if the price is too low, good contractors will be only attracted to work through the certainty that the contract will lead to further work for which realistic prices can later be charged;
- if the price is too low, there is likely to be a knock-on effect of dissatisfaction on the part of all those who are to use the finished facility in the long-term;
- the higher the price, the greater the client's perceptions and expectations of the quality of finished facility, service and support from the contractor;
- the higher the price, the greater the level of perceived demand that the client may be expected to make on the contractor.

Methods of payment

Virtually the only universal factor across the industry is that a proportion of the price is paid only upon satisfactory completion of the work or contract. This proportion varies from 30% to 100% according to the nature of the work, the provisions of the contract and client expectations. Some illustrations are given in Summary Box 4.3 and there are also implications for this in Summary Box 4.6.

The received wisdom is that a deposit or down-payment engages the client's commitment and also places a value on the work and on the contractor. Retention of a proportion also ensures that the work is done to the client's satisfaction. Staged payments may also be included – at their most positive, they reflect the continuing mutual commitment; and some contractors include them in order to ease cashflow problems when they are able (see Summary Box 4.3). Increasingly, the latter is being seen as an obligation of being in the industry in the first place across all its sectors, and clients are less happy to do this for this reason.

SUMMARY BOX 4.3 **Staged Payments: Examples**

Description	Initial payment	Stages	Final payment
Architecture and design	25% upon agreement	25% of modification (possible)	50%–75% on receipt of approved design
Town, country and environment planning	25%–50% at commission		50% upon acceptance of report/design
Double glazing	33% at commission	33% when work starts (possible)	33%–66% upon completion to satisfaction
Design and build	(a) 25% (b) None	May range from 0% to 50%	(a) 75% upon delivery (b) 100% upon delivery
Civil engineering	10%	(Possible) 10% at each of up to four stages (according to nature of contract)	50%–90%
Private finance initiative and other partnerships	Nil – down-payment only	Nil	Costs recouped through servicing finished facility
Quantity surveying	Free initial consultation	(a) Nil (b) Engagement fee in some circumstances (c) 10%–50%	(a) On completion (b) Balance (c) 50%–90%
Consultancy	Free initial consultation	(a) Nil (b) 10%–50%	(a) On completion (b) 50%–90%
Agency work	25% on commission	25% on appointment/interim report/acceptance of work of agent	50% – (a) after 6 months' satisfactory work, (b) on satisfactory completion of work

Relationship with demand

Organisations that have too much work chasing too few of their own resources, find themselves under pressure to put up their prices. This is fair enough, though there are caveats. The most important is client knowledge and understanding. It may be that the current client base only exists because of the current price, quality and value offering and that, if one element changes, the client may go elsewhere.

On the other hand, putting up the price may attract clients who perceive that they are now going to get enhanced quality and value as the result; or it may open up whole new sectors that were previously not considered by particular contractors because the prices were perceived to be too low to give quality of service. In the short-term, this may lead

SUMMARY BOX 4.4 **Differential Pricing: Examples**

- **Double glazing**: any seven windows for £995; buy seven windows and two doors included free; have the work done now and nothing to pay for 3/6/12 months.
- **Domestic products**: kitchen/bathroom/bedroom fittings – buy the units, free fitting; price levels include fitting in 7/28 days.
- **Building products**: brick, sand, cement – buy the products, free delivery; bonuses for early delivery; flexibility on contract work and just-in-time arrangements.
- **Civil engineering**: project delivery to include land and environment restoration, infrastructure, access; bonuses for early completion; may also include a commitment to hire a certain proportion of local labour.
- **Speculative house building**: houses to include carpets, curtains, kitchen, other furnishings free and to choice.
- **Speculative office/factory/industrial estate development**: facilities to include carpets, heating, lighting, telecommunications, other fitting out on demand and free.

In these cases the prices are used as part of the promotional and presentation effort. It is a common feature of consumer goods marketing and this indicates that industrial and commercial marketers are now looking much more widely at commercial pricing concepts.

to enhanced expectations; in the long-term, it may lead to dissatisfaction on the part of the new client or potential client base when the contractor finds itself unable to satisfy the increased level of demand.

Above all, this underlines the need to ensure that market knowledge and understanding are kept completely up-to-date so that any changes in expectations or sensitivity to price change are fully understood before changes are made.

The converse is also true – an organisation that has too little work is likely to be under pressure to bring its prices down. This is only sustainable if: (a) the lower levels of price reflect a long-term sustainable, viable level of business; and (b) if that is what the market wants. It may be, for example, that a market does not want lower prices but rather enhanced quality or attention to deadlines.

Again, the new levels of price may attract new customers and clients, and open up new sectors that have then to be satisfied. And, further, inability to do this will cause medium to long-term dissatisfaction – and therefore the price reduction becomes unsuccessful.

Price stability

Within their own constraints, clients expect price stability. They also expect contractors to be able to do the work to the agreed levels of quality and satisfaction once a price has been agreed.

It is true that contractors may go in at low prices, either to penetrate a new market or because it is better and more cost effective to work at a loss than not to work at all. Having said that, client bases do not expect wild price variations and these can and do lead to discontent and dissatisfaction. Especially among those who have been charged high prices, there is dissatisfaction once these clients get to hear that the contractor is prepared to do the work at a lower price elsewhere. Also, if a contractor carries out an initial project at a low price in the certainty/expectation that future work will attract higher prices, this should at least be understood if not specified. Such agreements are made between organisations – the negotiations are carried out by people. People change and, when this happens, the new individual may have neither the knowledge nor the inclination to honour anything that is not specifically contracted.

Target pricing

Target pricing is possible as the result of complete market knowledge. It enables contractors to package and present their products, services and expertise in ways familiar to the target market – and to price accordingly. This again is a form of differentiation and also enables one way around any problems caused by perceived price instability and variations among the market at large.

Geographical and regional factors

These have to be taken into account when proposing to work in particular areas and regions. Such factors include: levels of general prosperity; levels and nature of economic activity; the ability of the area to attract government and other public finance; the willingness of those in authority – both public sector and commercial contractors – to pay particular charge levels; and general levels of demand in the region or area for the particular product, services and expertise. It is also necessary to assess the availability of skilled labour in the particular area – if this is not present in sufficient quantities, then it will have to be brought in, and costed into the project. In some cases, for example when working in isolated areas, unskilled labour may also have to be brought in for the duration of the work. Knowledge and understanding of these factors enable a realistic assessment of possible price levels to be made prior to choosing to work in the area.

Political pricing

This involves reference to specific national and local government priorities and enables again, a realistic view of the short, medium and long-term opportunities and returns.

This also means reference to specific lobbies and pressure groups. For example, it may be possible to bid for government and local government contracts at prices that politicians are prepared to pay but which are nevertheless unacceptable (especially unacceptably high) to the public at large. Less frequently, public bodies may seek to commission projects at prices that are so low that informed public opinion and other lobbies, especially mass and specialist media, pronounce that the finished facility cannot possibly be of any value.

Moral pricing

In price terms, moral dilemmas arise when:

- There is an overwhelming need for a specialist facility or service very quickly and the client is prepared to pay almost any price. The contractor's decision is then based on whether to overcharge vastly on the basis that it will never work with the client again or to resist the temptation on the basis that it may be the introduction to a long and profitable relationship.
- To do the job at a good price might help the founding of a longer-term relationship.
- To do the work at a normal or reasonable price might be good publicity because the facility is important to the community, and to overcharge in this one case is certain to be seen as taking advantage of someone's specific pressure.

Parcel pricing

Parcel pricing occurs where several contracts are offered to, or won by, one contractor for a single overall price. Within the parcel there will invariably be individual elements that operate at a loss when seen in isolation but which are compensated for by the advantages of the other parts of the total volume of work. The advantage to the client is the ability to agree a variety of work with a single contractor; the advantage to the contractor is the ability to develop and demonstrate a variety of expertise, flexibility and customer responsiveness which may lead to extension of the working relationship.

Figure 4.6 is a useful indication of the relationship between the hard (income) and soft (presentation) factors that have to be addressed in establishing realistic and sustainable levels of price. It is useful to make the point that negative perception does not necessarily lead either to income loss or loss of work. Much work that has a negative connotation is nevertheless essential and regularly contracted (see examples in the figure).

There is a direct correlation, a clear relationship, between high levels

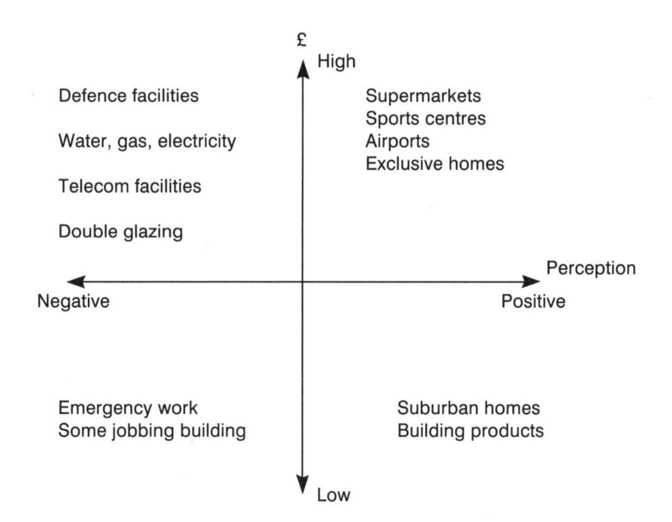

Figure 4.6 Price and perception.

of charge and positive perceptions; and the converse of this is that, when the positive anticipations and satisfaction levels are not forthcoming, there is a very quick decline in the quality of contractor–client relationship. This indicates one view of the concept of overcharging which is a client behaviour reaction where prices are high, but where there is no supporting perception of prestigious association or high quality of work. The position may be sustainable for some time by the contractor where substitutes and alternatives are not available. The position for the contractor can only become easier if it pays attention to the more positive perceptual issues.

Other aspects of price

The following points also need to be understood – they have specific applications to different parts of the industry:

- **List pricing:** akin to tagging and of a special significance to building products, domestic and commercial fittings and furbishment. List pricing gives clients the opportunity to browse and prospect and also enables other features such as design and installation to be demonstrated – for example, when presented as an illustrated catalogue or in a showroom. The weakness is that it gives the impression of inflexibility. This may cause buyers to be drawn to potential contractors and suppliers that give a greater impression of flexibility.
- **Discounting:** this is offered as a part of contractor–client relationships as follows –

(a) regular and agreed discounts for regularity, value and volume of business;
(b) as part of a broader pricing strategy, such as penetration or loss leadership, with a view to gaining a foothold with the new client or sector;
(c) for building products and domestic and commercial fixtures and fittings akin to seasonal retail sales, either to generate interest or cashflow; to improve trade volumes; or to clear old stocks in preparation for the new;
(d) as part of a pilot project or in return for other benefits – for example, a contractor may agree to build six houses to a particular design and price to see what happens in the market in return for a guaranteed higher level of price if the scheme is successful and leads to future work; or a contractor may carry out a public project free or very cheaply in return for a longer-term commitment to work provision;
(e) perceived discounting which is where the discount is used very much as the promotional effort, meaning that rarely, if ever, is the full price actually charged;
(f) where the contractor offers discounts in return for early settlements of accounts.

- **Trade pricing:** this occurs where organisations in a sector all set the same (or very similar) price levels. The strength of this approach is that everyone knows the sector price structure and this means that competition is on other factors, e.g. design, appearance, access, service and speed. The main weakness is that it leads to complacency. It should also be noted that price fixing agreements are illegal in the UK and the EU; they are considered to be a restraint of trade. This form of price fixing is called a cartel.

  It is also true that where trade pricing has existed for a long while, the industry becomes susceptible to new entrants who can produce the same or equivalent product/services to a greatly reduced cost base – and, therefore, with greater flexibility of price. This also happens to a lesser extent where one of the existing members suffers losses and takes unilateral action to break the trade price.
- **Skimming:** skimming is where there is an opportunity to take short and medium-term enhanced profits. The strength of the approach is in generating immediate and sometimes continuing income, and is quite legitimate provided that the sector is willing and able to sustain and support it. It has to be seen in the context that: it should never become a cause for softness or complacency just because enhanced levels can be generated; that it is certain to make the market attractive to potential entrants; and that volume, quality and service levels must still be improved and enhanced wherever possible be-

cause the client base will look around for alternatives and substitutes.

- **Expert pricing:** clients are prepared to pay well for expertise provided that the expertise delivers what is promised, indicated or inferred. High charges therefore indicate high levels of expertise and the converse is also true. Low prices indicate low levels of expertise. There are caveats to this – the actual price level: this has to be sustainable by the sector (e.g. there is no point in a consultant charging £200 per hour if nobody takes up the service); and it is a key in deciding to vary price levels in relationship to the specific sector served (e.g. the same consultancy may charge £200 an hour to financial institutions and £40 an hour to the voluntary sector).

  Conversely, it is no use charging £40 per hour to a sector with a view to gaining a foothold if the sector perceives that anyone who is any good will be charging £200 per hour. In this case, penetration pricing is likely to be in the order of £160 to £180 per hour and allied to other differentiated activities.
- **Price of other experts in the sector:** which may vary widely according to the differing nature of the expertise, i.e. planning, design, architecture, geotechnical drainage, structural engineering, quantity surveying, facilities management and project management consultancies may all be able to charge different fees according to their own expertise or may have to work at least within a broad framework.
- **Balance of power:** this refers to the ability of the contracting organisation to set a prescribed fee level and to attract the expertise of the required quality at the appropriate times to carry out work for it.
- **Price cuts:** price cutting is attractive to clients on the face of it, simply because they will be paying less for their products and services. The problems with price cutting are: that they may not be sustainable for any length of time, and this is certain to result in client disappointment; that everyone else in the sector may follow the leader which puts even more pressure on the first cutter, especially where that organisation has made the cut to give it a sustainable advantage; that the cut is allied to a decline in quality, volume and value of product, project and service.

  Cutting is therefore only sustainable in the long term where it follows or runs alongside improved cost efficiency and effectiveness.
- **Price rises:** there come points at which these have to take place. The issue therefore becomes, not that the price will go up, but when and by how much. It can be related to enhanced and improved quality and value of service provided that is what the client base wants. Large jumps in price may be made infrequently which tend to give the client base an initial jolt but then allow a new stability to develop. Price rises may be phased in gradually. They may be negotiated and agreed for continuing work with long-term clients and

LEEDS COLLEGE OF BUILDING

this is legitimate provided that the broader quality of relationship can be sustained and enhanced.

As indicated above, price rises may also be decided on in order to take the organisation upmarket and into higher-quality, higher-value work. In this case, the new levels of charges will be sufficiently high to attract a new client base to the organisation. However, this has clearly to be conducted alongside other marketing activities that underline and enhance the real and perceived quality of work that is to be carried out in the future.

PUBLIC SECTOR CONTRACT PRICING

This is to be seen in the long term for the following reasons:

- Where there are price constraints, where contracts are awarded overwhelmingly to the bid that offers the lower price and where a price is set for the work to which anyone wanting to do the work must exceed. The decision to take part is based on whether or not the contractor wants to be a public service contractor. If this is the case, it becomes one of the rules of engagement and the pressure is therefore placed on the particular organisation to structure itself in order to be able to do the work in this way and within these constraints.
- Where there are specific quality constraints that have to be accommodated within the price, such as those beginning to be used by HM Treasury and other central government departments.
- Where the cost of the work is borne entirely by the contractor, and only to be recovered from making charges to the end-users of the project or facility (e.g. as with private finance initiative projects).
- Where funding is not a problem, where there is plenty of public money, where the first priority is to get the particular facility built (almost regardless of cost). It is tempting (as shown earlier) to enhance the price, especially where recent previous work has been carried out under severe restraint. This has to be seen in the context again of deciding whether to be a long-term player in the sector and therefore as a step to securing the long-term future. When the work is heavily demand led there may be scope for the contractor to secure further work in return for reducing the price; if not, then the decision can be taken to do the work for the maximum possible price.
- There is scope for greater price flexibility in some circumstances. This occurs towards the end of public bodies' financial years when work has often to be commissioned quickly in order to use up resources that would otherwise be lost. It also happens when govern-

ment departments delegate resources to a local management body for a specific project (e.g. an NHS Trust delegating resources for a new hospital; Department of Transport using a motorway or rail contract as part of opening up or regenerating a depressed area), and this often includes part of the spin-off work that is generated in turn (e.g. other health trust work in the hospital case; other infrastructure work in the motorway or rail case).

It is also essential to recognise the threats and instability inherent in long-term pricing in seeking and securing public service contracts. There is so much outside the control of the contractor. At the top of the list comes the client's political pressures – as political control of public services changes, so does the attitude and approach to the industry and can include:

- whether or not to use local, regional or national contractors;
- whether or not to use foreign contractors;
- whether or not to place specific contract compliance constraints (e.g. the percentage of local labour to be used; training agreements; union recognition);
- the nature of the fund or budget from which the work is to be paid for, e.g. a UK government department budget may be heavily constrained, but if the work can be paid for by the lottery fund there is much less limit (e.g. the UK national stadium could not be funded by the Department of the Environment but is being paid for by the lottery fund which has plenty of money);
- whether or not to use in-house provision (especially important in local government and health service contracting);
- whether to engage in a partnership akin to the private finance initiative or other equivalent – whereby the contractor and client work in a joint venture or whereby the contractor underwrites and pays for all the work in return for being able to charge for its subsequent use.

Related to this is the propensity of politicians to change their mind and direction unilaterally and often without reference to, or consultation with, the contractors. A long-term relationship with the public body has therefore to be seen in a very broad context. While in the past, during periods of public sector growth, it was possible to be a mainstream long-term public service provider, such a point of view is not sustainable at present in the UK without reference to these constraints (see Summary Box 4.5).

SUMMARY BOX 4.5 Public Service Contracting Constraints

The following have also to be considered when contemplating working for – and pricing work for – public bodies:

- **Cancellation:** especially unilateral cancellation; requiring the broadest possible view of the political situation.
- **Moratoria:** considering the length of delays or contract freezes that may be engaged in and the reasons for these.
- **Short termism:** the ability to respond to short-term pressures, especially towards those indicated in the main text at year end.
- **Lobbies and vested interests:** which may heavily influence such things as media coverage of works; examples of these are illustrated elsewhere in the text.
- **Pressures from elsewhere:** which arise, for example, as the result of a politician perceiving that greater quality, expertise and value can be secured from Japanese/French/German/South African contractors.
- **Too many contractors chasing too little work:** very attractive from a political point of view, as it is perceived to enhance competition.
- **'If you don't want the work then there are plenty of companies that do':** this point of view is a mark of extreme disrespect and lack of value, and any future contractor–client relationship engaged in with bodies that exhibit this point of view has to be seen in this context.
- **Expertise–influence:** as stated elsewhere in the book, many public contracts are awarded by persons of great influence but little knowledge or expertise concerning the industry itself.

CONCLUSION

The key to effective pricing is therefore understanding. For any activity there is no absolute correct level of pricing or price fix other than that which will sustain the long-term security of profitable activity; and because of the current and envisaged nature of the industry, this has to be a matter of constant review. On the other hand, there is clearly an incorrect approach, and this is where prices are plucked out of the air or established as the result of perceptions and received wisdom only, rather than on a full analysis of the organisation's position and capability, and the capacity and willingness of the market to sustain particular charge levels.

Understanding is achieved as the result of considering each of the factors and elements indicated in this chapter. Where success is achieved, reviews can then be carried out to assess why the pricing policy was successful and to build on this for the future. Where failure was the outcome, where prices were set at the wrong level, this too can then be

SUMMARY BOX 4.6 **Pricing Differentials**

One very famous geotechnical consultancy is in the habit of charging a fee of £2600 per day to its clients. It has just recently opened up a major new sector of work. The company is held in very high regard in this sector and, if it gets the price right, it will secure a highly profitable range of activities that the particular company can easily sustain.

The sticking point is on price. The new sector is used to paying between £1000 and £1500 per day for these services. The company is therefore in something of a quandary: whether to stick to its normal pricing policy for the new sector and take a bit of a chance that their other qualities will secure the work in any case; or whether to reduce its fees in order to meet market expectations. If it does this, it may face specific questions from its current range of clients about their existing pricing levels.

The option of keeping the matter a secret would, by common consent, not be viable. People move regularly around the sector, and charge levels – together with other business practices – are common sectoral knowledge.

There is no right answer. A company will decide from its own justified point of view the right approach to take. It is however a useful illustration of the problems of differential pricing. The strength is that it gives the opportunity to maximise and optimise resources, to break into new areas of work, and to match price with the combination of demand and expectation. The weakness is that those who are charged the higher rate inevitably come to wonder whether they are being overcharged for a product or service that is self-evidently available at lower rates.

worked back to assess where and why mistakes were made so that these can be remedied for the future.

The final point to make is to ensure that price is seen in the context of the complete marketing mix. It is the meeting point of contractor and client and, from an economic point of view, the basis of agreement to work. Even when a contract is awarded purely on price however, the client has still to be satisfied that the contractor is able to carry out the work to the required quality and value, and this is actually based on knowledge and understanding of the other factors (whether stated or not) including reputation, past history, track record, known and perceived expertise and capacity.

5 Product

In marketing, product refers to the totality of the offering – and this usually includes services offered by the contractor in support of the physical product. A comprehensive definition is given by Philip Kotler as follows: *'A product is anything that can be offered to a market for attention, attraction, acquisition, use or consumption.'* (Kotler, 1993).

For the construction industry, products are:

- **Physical:** houses; public and commercial facilities; infrastructure; building products.
- **Services:** expertise; consultancy; design.
- **Persons and reputation:** Richard Rogers, Norman Foster.
- **Names:** McAlpine; Mowlem, Laing, Tarmac, Ove Arup; Everest; Jewson.
- **Ideas and proposals:** urban regeneration; regional development; competitions.
- **Essential:** water, gas, electricity, telecommunications, transport facilities.
- **Highly desirable:** schools, hospitals, shopping and leisure centres.
- **Desirable:** environmentally secure; adaptable.

PRODUCT LEVELS

The following product levels may be distinguished.

- **Primary benefit:** identifying the product's primary function. For example, the primary purpose of building a dam is to provide water for a specific area; and the commissioner of the dam project is buying a water supply.
- **Generic product:** this is a more general definition of the area of expertise. As examples, civil engineering is concerned with combining steel stock, concrete and other primary materials into facilities. Design is the production of the client's vision of the future. Consultancy is problem solving.
- **Expected product:** this relates the product to the client's expectations. It is a set of attributes and qualities normally expected by the client when the work is commissioned. For example, completion of a road contract is expected to sustain traffic flows in the medium to long term. A sports centre is expected to include swimming pools, games facilities, recreation and leisure areas. A house is expected to be cool in summer, warm in winter. Problems solved by consultants

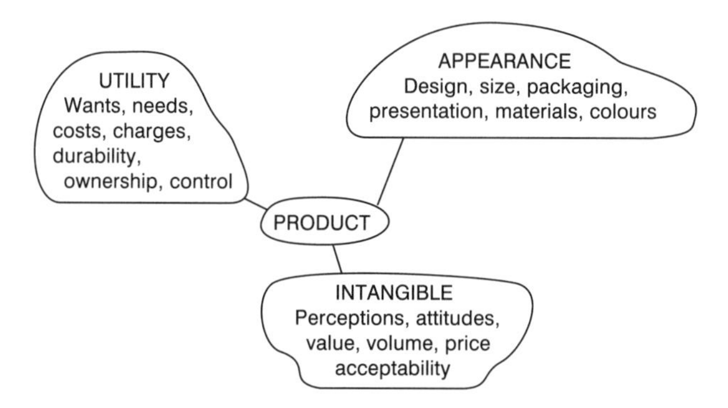

Figure 5.1 Product features.

are expected to take account of foreseeable future circumstances, rather than substituting one set of issues for another.

- **Differentiated product:** this is where the product is distinguished from those offered by other players. For example, a double glazing company includes free fitting or security locking; a building products company offers free home or site delivery; a civil engineering company undertakes to use a percentage of local labour for the duration of its activities.

  In the domestic and commercial building sectors, differentiation is being heavily developed by some companies; and includes the installation, servicing and staffing of safety and security systems; maintenance; furbishment, refurbishment and upgrading; and transport access.
- **Product potential:** this is the fullest consideration of all the uses to which the product may possibly be put in the future. For example, in the future, a multistorey car park may possibly be transformed into offices or factory units; office and other commercial construction may possibly be transformed into domestic accommodation in the future.
- **Product identity:** this is the production of something to the client's satisfaction which also meets the contractor's needs to be identified and associated with a long-term enduring and highly successful project (see Summary Box 5.1).
- **Product integrity:** this is the sum of all the features and properties of the product and related to the purpose for which the client has purchased it. It is a combination of total product performance with a full understanding of the total effects on the client's future well-being, the effects on the end-users and wider general public and professional perception and reception (see Summary Box 5.2 and Figure 5.1).

SUMMARY BOX 5.1 **Product Identity**

In the UK and much of the Western world, the building and construction industry has traditionally been product rather than client led. In this context, the consequence has been a tendency to present the contractor's distinctive expertise and format as available for choice and consumption rather than by the client base; the physical products rather than the expertise and capability were the driving force.

In situations where there is too much work chasing too few providers this is less critical, at least in the short to medium-term.

From a marketing point of view however, the product has to meet the client need rather than contractor willingness or familiarity. The output must be that product service or finished project first meets the client's needs. Any specific need on the part of the contractor for identity and recognition has to be harmonised with this.

Example: A38 river crossing at Plymouth

This project, undertaken in late 1993, is a shining example of how to meet both contractor and client product needs.

The A38 road bridge across the River Tamar needed replacing. The traditional approach would have been to have closed the bridge off, dismantled it and erected a new structure. In the normal course of events, this would have involved the road being closed for several months; traffic would have had to have been re-routed, causing considerable congestion elsewhere.

Following consultation between the client (Department of Transport) and the contractor (Balfour Beatty) a different approach was taken. Balfour Beatty built the new bridge at the side of the River Tamar. When it was completed, they arranged with the client to close the road down for one weekend only. Huge cranes were put in place; and these first severed and lifted the old bridge away from the existing road and were then used to install the new bridge in place.

Everything worked according to schedule. And – more importantly – both contractor and client needs were met. The client received the finished facility, capable of sustaining traffic flows for the foreseeable future; and this was achieved with three days' disruption only. The contractor received widespread coverage, both national media and professional press, for its expertise; and the bridge now stands as a monument to this achievement.

SUMMARY BOX 5.2 **Product Integrity: Compromises**

The following are examples of where product integrity has been compromised.

The Great Wall of Hulme

This was a housing development constructed in the 1960s on the western edge of the city of Manchester. It was designed to rehouse up to 30,000 people as part of a wider programme aimed at transforming the poor residential parts of that city. Those who came to live in the Great Wall were to be removed from the houses built in the 19th Century, endless terraces of back-to-back and wall-to-wall basic housing.

The Great Wall was so called because of its appearance and structure. It consisted of four blocks of 3000 two and three-bedroomed flats and maisonettes. Each was ten storeys high and 1000 yards long.

From the client's point of view it had the following attributes: it was good value (cheap); it would be straightforward to build; it would satisfy the declared political aim of rehousing a large number of people; it would release land for other activities and amenities; and it was part of a wider plan (that of transforming the city of Manchester).

From the contractor's point of view it provided a substantial amount of work and (if successful) the chance to become the leading exponent of this kind of development. It brought extensive furbishment, furnishing and house-fitting work, and subcontract infrastructure (water, electricity, telecommunications and access).

The users' point of view was never considered. There was no understanding of life in the long-term in such a facility. No amenities, living and leisure facilities, or factors concerning the general quality of the environment were ever provided. This was partly because the client subsequently became unwilling to invest further in the project, partly because it became clear during construction and from academic and industrial studies that what was being created was a social and environmental disaster. Notwithstanding, the project was completed and people were duly housed in the blocks. The facility had a useful (if that is the word) life of 14 years, after which it was demolished.

The Pergau Dam

The Pergau Dam was commissioned in 1985 and constructed during 1987–91. The stated purpose of this project was to secure the water supply to a large part of central Malaysia. It was arranged with the support of both the UK and Malay governments, and was subsidised by the UK government as part of its overseas aid project.

It ran into trouble for four reasons. The first of these was the considerable negative impact on the environment where it was built. The second was that it did not do the job for which it was commissioned – its impact on the security and quality of water supply was negligible. Thirdly, the price of the project rose considerably and this led to an argument between the contractor and the governments of the UK and Malaysia over who should pay the difference. The fourth – and most notorious – reason was that it was tied into an armaments deal between the UK and Malay governments – the dam would only be built if the Malay government took a particular volume of military hardware at the same time.

Special treatment

This is concerned with product delivery or project completion. In the marketing of consumer goods *'special treatment'* is that which *'delights'* the customer. It consists of benefits such as: free teletext on all TV sales; free packet of video tapes with video recorders; free reception with the sale of a wedding holiday package. Provided such benefits come as extra, are targeted at the anticipated needs of the customer and are of the same high quality as the product itself, they are highly positive, highly distinctive and highly desirable benefits.

In construction, examples of special treatment are:

- **Builders:** garden layouts; carpeting and curtains; kitchen, bathroom and bedroom furnishings.
- **Civil engineering:** site refurbishment and replacement; staff training; hiring local labour; giving access to schools and colleges; provision of landscaping/open space.
- **Architecture and design:** choice of designs; early delivery; extensive customer liaison.
- **Building products:** instruction in how-to-use (domestic); precise guaranteed delivery times (industrial and commercial); after-sales service.
- **Consultancy:** a range of answers and options plus full briefing on each; maintenance manuals; cost–benefit studies.

PRODUCT CONFLICT

The demands and needs of the contractors, clients and end-users of any facility or project are different and have to be reconciled. Where this is not achieved, product conflict exists.

Architects produce designs to enhance their reputation for quality and innovation; to make a distinctive mark; to secure their next commission. They may also have specific targets to work to – for example, if the design commission is to be secured by competition, they have to satisfy judges who may be influenced by the need for controversy, distinctiveness – and, therefore again, enhanced reputation.

Builders/contractors and civil engineers/architects produce the finished item and have to deliver to the satisfaction of the client. This always means reconciling design and appearance with materials usage, facility requirements, durability, longevity and potential flexibility. They also have an eye on producing something that is a monument to their innovation and expertise – and which, again, may influence potential clients.

The clients have first to reconcile what is within their control and what is not. Those who commission public projects are invariably constrained by cash restraints, often imposed by government and over which they have little or no influence. They may be obliged to take the winning design from a competition – whether or not they want it. They may (and do) often have no particular knowledge of the matter – only that they are charged with the responsibility of commissioning it. They may (and do) have political, social and environmental considerations, and have to respond to and accommodate these pressures. Clients may also expect after-sales service, facilities management and flexibility of usage as part of the product (see Figure 5.2).

There is, therefore, immediate potential for conflict and this can lead to a loss of focus on the part of each or all of these players (see Summary Box 5.3). Clearly, the greater the clarity established at the outset, the greater the likelihood of successful activities.

SUMMARY BOX 5.3 **Product Conflict: Other Players**

Other players also have their own views. National and professional institutions, such as the National House Building Council, the Chartered Institute of Builders, the Royal Institute of British Architects, the Institution of Civil Engineers and the Royal Institution of Chartered Surveyors, expect to see work that is exciting and innovative, and which also meets the quality and value standards that they set for the industry as a whole. They have a general role in promoting the interests of the industry; and may also have to become involved in reconciling the differences of interest indicated above.

- **Quantity surveyors:** the product of quantity surveyors is their accuracy and attention to detail. This concerns their estimating and delivery expertise, materials quality and access to sources, cost management, construction procurement and contractual matters.
- **Consultants:** the product of consultants is their expertise. This includes relating to the specific matters in hand; and also personal and professional qualities of flexibility, creativity, innovation and quality of problem solving.
- **Planners:** the product of the planner is the environment created at the end of activities. This includes expert public and media perception and acceptance, as well as the 'hard' use of the environment and quality of life achieved.
- **Subcontractors and agencies:** their product is satisfying the main contractor. This includes the presence and application of their expertise; in many cases, this also includes their capacity for problem solving. It has reference also to the speed with which they are available and the quality of work that is produced.

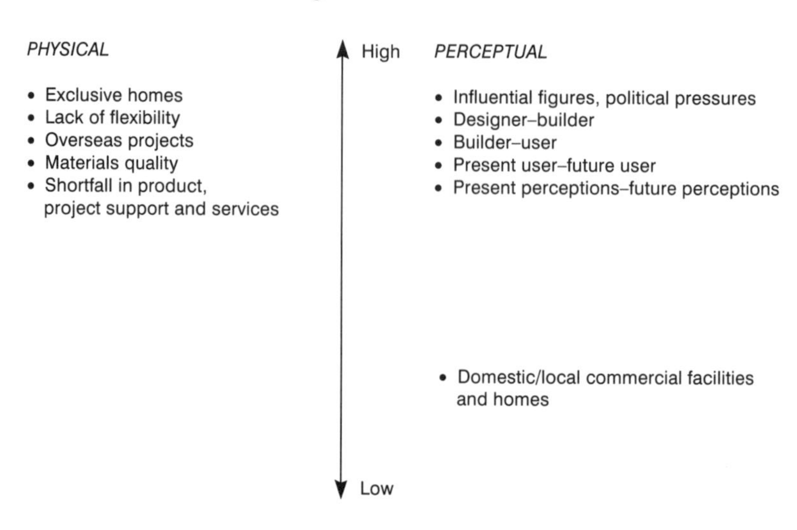

Figure 5.2 Product conflict potential.

Other features

Other features of the construction industry product are:

- **Environmental effect:** new or proposed initiatives have to be reconciled with the existing built and rural environment. They have to be presented in terms that acknowledge this and re-emphasise the benefits to the locality and its community.
- **Specific blight:** this refers to specific problems – traffic and pedestrian re-routing and diversions, dust, site access and delivery – and their effects, especially on those things important to the particular locality and its people. Continuing domestic and existing facilities access have especially to be considered.
- **Public acceptance:** this means viewing the product in terms of its uses to those directly affected and also to anyone else who may be affected in the future.
- **After-sales:** this is more straightforward where there are industrial standards, legal guidelines and guarantees in place (e.g. National House Building Council (NHBC), Glass and Glazing Federation (GGF)). It is less straightforward when –
  (a) an after-sales agreement is not specified in the contract;
  (b) an after-sales agreement is specified in the contract but does not pin down specific responsibilities (e.g. is a subsequent problem concerning a drain the responsibility of the main contractor or the drainage subcontractor?);

(c) an after-sales agreement does pin down specific responsibilities, but where those who carried out the particular work are hard to trace and engage;
(d) specific responsibilities are pinned down, but where the problem does not fit easily into one of the specified areas (i.e. the unforeseen circumstance);
(e) specific responsibilities are pinned down but no future timescale is given.

- **Long-term effects:** organisations that build for one era may be welcomed with open arms as providers of work, housing, amenities and facilities. As public perceptions and expectations change, all this is forgotten and the company is remembered only as being entirely responsible for something that has degenerated into blight and environmental destruction.
- **Reconciling the product with its environment:** this occurs, for example, where a contractor is commissioned to build a hospital in an area known for its lack of easy access. The first problems are traffic congestion caused by the contractor's requirement for workforce and site access, deliveries and plant usage. Subsequent problems are apparent when the finished facility comes into use – and for a hospital requiring absolute speed of access in case of emergencies, this is potentially very serious. Any potential clash, conflict or incompatibility of usage with adjoining or neighbouring facilities has therefore to be prescreened effectively.

  Architects and planners may be asked by clients and potential clients to draw up plans for particular schemes, knowing that they will create a form of domestic and environmental blight once it becomes known that proposals are in hand – and that if they do not do this work, it will nevertheless be given to someone else. This is most common when bypass and motorway schemes are being mooted. If not handled properly, house and land values in the vicinity can collapse; while land values along the proposed route may sharply increase (see Summary Box 5.4).

- **Perceived benefits:** benefits accruing from a project that are obvious to a client or direct user are not always apparent to non-users. The benefits of a project have not only to be apparent, they have to be known and understood to be apparent by everyone who has any form or level of involvement or interest. Fundamentally, rationally sound projects and products fail because positive attitude, image and response have not been engaged, not because the project itself has no value or merit.

SUMMARY BOX 5.4 Reconciling the Product with its Environment: the M20 and the Channel Tunnel Rail Link in Kent

Between Ashford and Maidstone, the route chosen for both of these projects was largely the same. However, perceptions and acceptance of each was very different.

The Channel Tunnel rail link (as documented elsewhere) was the subject of long and bitter campaigns against it – and these campaigns were based on facts, half-truths, myths and legends surrounding: noise levels of the trains that would be using the link; general environmental blight; opposition to the Channel Tunnel itself; and environmental destruction. On the other hand, the M20 went through by common consent and without opposition, in spite of the fact that the finished facility produces: pollution caused by traffic fumes, noise and lighting; property price blight for those who live near to the motorway; and the 'eternity' of the motorway itself.

From a marketing point of view, the acceptance of the one and the rejection of the other is down to the relative ability of both contractor and client to satisfy the perceptions of those affected by the value of the finished facility. The motorway was perceived to bring benefits of convenience and access that outweighed the blight. No such benefits were made apparent concerning the rail link – and this is in spite of the fact that noise and pollution levels would not be nearly as high as those generated by the motorway; that work would be brought into the area at least during construction; and that the finished facility would open up opportunities for long-term work and economic improvement along the route once it was completed.

- **Political factors:** these occur most commonly where –
  - (a) whatever is proposed is fundamentally sound but to do it to the required scale or standard would cause a public outcry against it: this occurs frequently with new road schemes, especially motorways; with low to medium quality mass housing development; and with the creation of new towns and cities in rural areas;
  - (b) what is proposed has less merit than the political drive to create work and take steps to rejuvenate an area badly hit by the loss of its economic base (e.g. the close of a main/dominant industry): work for the construction industry includes roads, town centre refurbishment and shopping centres, car parks, sports and leisure amenities, and industrial estates – sometimes alongside political lobbying elsewhere to get companies to move into the area; these schemes have least merit when the economic activities created – largely retail and services based – cannot then be sustained by the area once the construction industry finishes its work and moves out;

SUMMARY BOX 5.5 Building the M25: In Three (At Least) Stages

The M25 motorway is the London orbital traffic route. It was conceived in the early 1970s. At the outset it was envisaged that it would take national traffic away from the centre of London, thus relieving congestion. Everyone would benefit – through traffic would not get caught up in urban traffic jams; and urban traffic jams would be removed because of the absence of through traffic. Frequent different points of entry and exit were designed into the scheme so that through and long-distance traffic could be accommodated from anywhere in the UK.

When the project was first modelled though, it quickly became apparent to those with political and other vested interests that the scale of the project required to do the job properly would be unacceptable. Nor was sufficient account taken of the fact that people would use the motorway for short stretches of travel (e.g. coming on at Junction 1 and going off at Junction 2 or 3) – that is, the motorway would attract domestic and local as well as through, national and long-distance traffic.

The result of these pressures was that the motorway is grossly overloaded and had to be – and continues to be – constantly reconstructed. Originally planned as no more than a two-lane dual carriageway, it quickly had to be upgraded to a three-lane highway; it is now envisaged that eventually, in order to be fully effective, it will have to become at least a six-lane super-highway – and that because of the prevailing political climate, this will have to be carried out in stages rather than as a one-off mega project. Moreover, the facility is constantly having to be upgraded in order to take the ever-increasing volumes of traffic that use it.

In this case, the overwhelming pressures have been a combination of perceived benefits and political factors. While the prevailing attitude to the M25 is negative, the perceived benefit that it offers far outweighs this and people therefore use it. In political terms, the construction of an adequate facility on a one-off basis was unacceptable – and the current approach is deemed to be that which is the least unacceptable in the circumstances.

(c) what is proposed is 'distributive' – which means that, as the result of the creation of the new facility, existing facilities suffer; a common example of this is the negative effect (both real and perceived) of edge-of-town superstores and shopping malls on existing town centre shops and amenities.

This represents the total context in which the physical features of, and other attributes of, the product become commercially viable. Attention to these features then concerns:

- **Design and appearance:** which has to be a combination of distinctive, durable (rather than faddish), acceptable (again, a very nebulous idea) and meeting the client's and users' expectations (for example, there are clear perceptions of what constitutes the design expectations of a road bridge, an executive home, an office block, a factory – without always being precisely able to articulate what these are).

- **Longevity and durability:** these refer to the materials used and the ways in which they are combined. Again, expectations as well as performance have to be met – houses are made out of bricks or stone rather than concrete or steel; and if concrete or steel are used, then a cladding is required. The materials themselves must also have their own basic integrity and this should apply whatever the attitudes of the client or nature of the relationship (see Summary Box 5.6).

SUMMARY BOX 5.6 **Materials Durability: The Palermo Underground Transit System**

The Palermo Underground Transit System was built in 1989. Initially it was provided for the great influx of tourists and visitors to the city of Palermo for the football world cup held in Italy in 1990. After that, it was anticipated that it would have great future use potential – during the world cup, both the city of Palermo and also the whole island of Sicily would receive favourable television and media coverage, and this was perceived to have great potential for the tourist industry. Moreover, the notorious crime and Mafia problem was being cleaned up at the time and the perception was that people would be coming to a crime-free city and island.

Within a short time, the stations and rail network started to crumble and the whole system rapidly fell into disrepair and disrepute. While top standard materials were specified, the contractors actually used substandard materials, allegedly with the connivance of the client (the city of Palermo), so that the savings in price could be shared out among those involved. The end result is that, in order to resolve the problem finally, the work will have to be done again virtually in its entirety.

- **Utility:** the product must bring the benefits for which it was designed and envisaged. There are variations on this: the benefits may exceed those envisaged; or it may bring a range of benefits not originally envisaged but which nevertheless actually satisfy the needs of the customers and clients (see Summary Box 5.7); or it is quickly realised that a project is a mistake, but which then leads to the matter being put right ('right second time'). Where none of these exist, neither does utility.
- **Presentability:** on the question of presentation, there is no problem with products that have physical existence and distinctive identity. For some design, architecture, planning and major projects and construction activities, this can also be addressed (in part at least) through

SUMMARY BOX 5.7	**Utility: Double Glazing**

Double glazing is often bought by both domestic and commercial customers and clients with the purpose of enhancing property values and reducing heating bills. This does not always occur. However, derived benefits include greater security that may, in turn, lead to reduced insurance premiums. It also eliminates condensation and reduces sound transmission. So, while not necessarily enhancing the price, double glazing may nevertheless make the facility more instantly attractive at the original price; and it therefore sells more quickly than equivalent facilities that do not have double glazing.

the production of mock-ups and models and through reference to existing near equivalents elsewhere.

The problem is more serious where this is not possible, and where the service offered is purely problem solving, or a project management or specialist agency. Again, reference to near equivalent assignments is valuably supported, where possible, by site visits, photographs and other presentations. Ultimately, however, the 'hard' features of design and appearance, longevity and durability, and utility have to be satisfied through incorporating their key elements in the overall organisation presentation and the expertise of direct and indirect contacts, as part of the process of building an effective and profitable contractor–client relationship, so that these are ultimately covered through the direct past and recent history of the relationship between the two. If marketing to clients where there is no past history of professional involvement, this equivalent of the product has to be capable of presentation in a variety of ways that meet both the client needs and perceptions, offer a substantial service of integrity and present the track record gained elsewhere, so that as full a picture as possible of the particular expertise is given.

PRODUCT MIXES

Product mixes are made up of the total range of activities offered for sale to the market. The basic ways of looking at these are:

- core and peripheral products;
- product portfolios and life cycles;
- branding;
- new product development.

Core and peripheral products

Core business

For all players, identifying this is essential – and it may not always be easy or straightforward (see Summary Box 5.8). It is either:

- the work that provides the majority volume of activities;
- that which provides the greatest volume of income;
- that which provides the greatest indication and enhancement of reputation, confidence, capability and creativity, and flexibility.

This may be restated as:

- products and activities that attract clients;
- products and activities that clients actually buy;
- products and activities that make money.

Peripheral business

The starting point for all successful peripheral activities is the drive to maximise and optimise the existing resources and cost base. This comes in the following main forms:

- Use of existing expertise in compatible fields, e.g. architects and building designers may get involved in consumer product design; civil engineering companies may develop a range of expertise, such as geotechnical services, water expertise and traffic management, to be offered on projects other than their own.
- Offering in-house provisions to outsiders: common examples of this are the development of management and technical training activities for offer on a subsidised or commercial basis elsewhere; the same sometimes applies to pension schemes and payroll; and to information services and libraries.
- Use of existing technology to produce products for other markets: this applies both to manufacturing and information technology. This is well established in the building products sector for example, where plastic extruded guttering and fittings companies can easily adapt to produce domestic and commercial packaging, toy and game components, and elements used in the assembly of domestic appliances. Company information systems can even more easily be offered on a commercial, subsidised or almost retail basis to outsiders.
- Response to additional requests from existing clients, either during the course of existing contracts or apart from them. For example, a civil engineering company that has undertaken to use a particular percentage of local labour may be asked by the client to provide

SUMMARY BOX 5.8 **Core Business Examples**

Architects

- Attraction/reputation: entering and winning competitions; design commissions for prestige futuristic projects.
- Sell: refurbishment; seminar and publication work.
- Make money: supermarket and car park designs.

Building/civil engineering

- Attraction/reputation: global perception (from annual report); domestic infrastructure projects; tendering for no-hope jobs.
- Sell: overseas projects; domestic high reputation projects.
- Make money: motorway and highway repairs.

Building products

- Attraction/reputation: dream homes, show homes; magazine and media features.
- Sell: commercial sales; also domestic sales, refurbishment, home improvements.
- Make money: national and local government, National Health Service, refurbishment.

Planning

- Attraction/reputation: published proposals for land and regional rejuvenation and restoration.
- Sell: high profile, innovative, reclamation and rejuvenation completed projects.
- Make money: continued consultancy and retention by public bodies.

Jobbing builders

- Attraction/reputation: private small good jobs well done.
- Sell: attitude; speed and flexibility of response; extras such as painting and decorating, kitchen and bathroom fitting.
- Make money: regular subcontract work for larger organisations.

Double glazing

- Attraction/reputation: variety, flexibility, different designs, Georgian bars, security.
- Sell: basic product, extras such as doors and weatherboarding.
- Make money: finance plans.

continuous training – both for the job in hand and also for the future – as an additional feature. In this kind of case, the main problem lies in ensuring that the quality of the peripheral activity is just as high as that of the main contract or core business.

- New product development as the result of known or perceived client and market expectations and changes. This often comes from the knowledge that work has been lost to a competitor on a specific issue and, rather than bringing in consultants, the company seeks the solution in-house.

As well as striving to maximise the resource and cost base, the benefits of peripheral activities are:

- Entry into, and familiarity with, new areas of activity that can then be assessed in terms of future potential and the implications of developing them further.
- Enhanced reputation, developing the perceptions among the client base that 'this is a top quality all-round organisation'; 'nothing is too much trouble'; and 'nothing is beyond the means of this contractor'.
- New product development, the opportunity to pursue notions and inclinations in what is at least a semicommercial setting.
- The derived generation of an enhanced commercial awareness among all staff. A contribution to the continuous assessment of the company's strengths and weaknesses.
- A contribution to the general drive for constant improvement, product and service development, meeting customer needs and wants.
- A form of subsidy for prospecting; while it remains true that marketing is a direct, core and priority activity, peripheral activities may be developed so that at the very least they contribute to the company's total fund of knowledge, understanding and expertise; and in many cases, this leads to general long-term commercial opportunities that can then be capitalised on.

The rules that should govern peripheral activities are as follows:

- They should remain peripheral unless, and until, the case for their full commercialisation becomes at least strongly indicated. The reason for this is quite simple – if and when they do become more fully commercialised, they are either going to divert resources away from the existing core business or they are going to add to the cost of the investment base.
- They should be financed from the existing resource and capital base unless, and until, it becomes apparent that additional resources, investments and expertise are required so that a full assessment of the genuineness of the opportunity can be made.
- Because they are fresh and adventurous, sight of the core business may be lost or blurred if the peripheral activity is not kept firmly in context. At worst, they may become glamorous and attract key members of staff away from their priorities and core activities.
- Transformation from peripheral to core business always requires substantial investment even when the case is proven absolutely. This investment means new staff, expertise, technology, information and

continued long-term access to substantial client bases. It is also certain that if the particular area is the peripheral business of the given organisation, it is the core business of others – and there is a world of difference between picking up odds and ends that are viable in themselves and developing substantial viable market shares in new fields.

- Extensive involvement in peripheral business may dilute effort – and just as important, reputation and confidence – in the existing client base.

The point is therefore to ensure that the aim of maximising resources is followed without diluting the core effort. It is also true that the peripheral business may only be possible because of the strength of the core business and consequent high levels of reputation and confidence.

Product portfolios and life cycles

Product life cycles

The useful way of looking at an organisation's products is to see them in terms of the life cycle. The general hypothesis is that all products have a finite life and that, within this, they are born, they grow, they develop and mature, finally reaching old age and eventually obsolescence.

In most cases, however, this process is more complex. It takes a great deal of invention, research and new product development activity to bring an item into being, and for each that is successful there are many that fail. Assuming that the product is intrinsically useful and valuable to a range of clients in some way, the life cycle may be influenced at each stage. At the point of youth and adolescence, for example, growth may be accelerated by effective marketing – and conversely, growth may be retarded or the product killed off altogether if the marketing activities are poor, ineffective or inappropriate (see Figure 5.3). Also changes of fashion and land use planning may render a building obsolescent before the end of its structural life.

At the point of maturity, marketing activity is conducted again for the purpose of gaining access to as many different sectors and segments as possible: for example, moving into new regions; repackaging well-tried and high-valued expertise; and going international.

At maturity and old age, products may be regenerated or rejuvenated – or killed off. Where regeneration and rejuvenation are considered, this is again likely to be achieved through repackaging, re-presentation or the offering of a range of new benefits and service support not hitherto considered.

Some products and offerings give the appearance almost of eternity. However, the one common factor to the success of all of them is the

LEEDS COLLEGE OF BUILDING

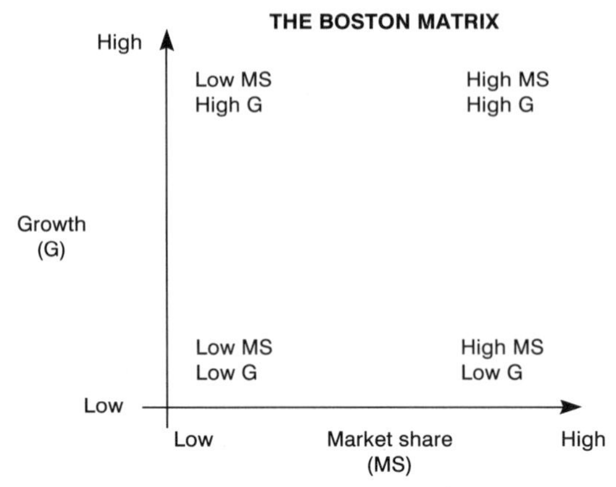

The model assumes that organisations have a range of products or offerings, and that these will be at different stages at a given point in time.

There are four quadrants in the model, for the assessment of products, and the basis for taking marketing decisions in relation to them.

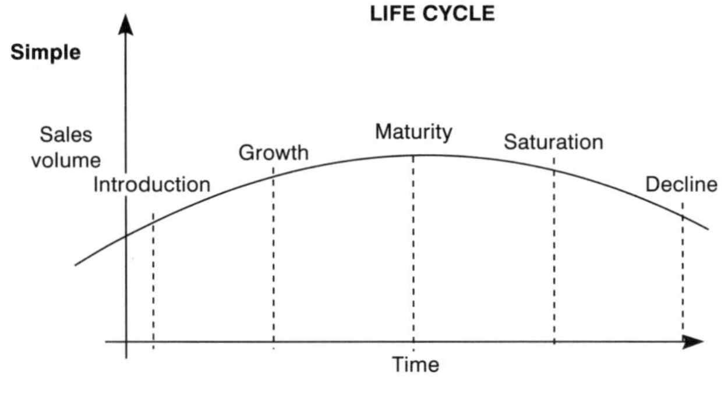

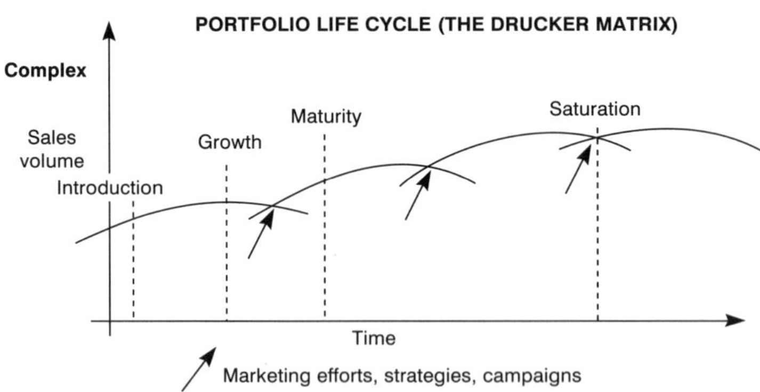

Purpose: the model indicates the regeneration, renewal and extension of the profitable life of the product concerned and its relationship with successful marketing activity.

Figure 5.3 Product life cycles. (Source: R. Pettinger, *Introduction to Management*, Macmillan Press, Basingstoke, 1994.)

continued ability of the company in question to re-present, rejuvenate and strengthen any images or brand features that are apparent.

Some products also have a decline phase. This may be for a variety of reasons. Examples include loss of competitive advantage over, and vulnerability to, substitutes. Technological advance may render a particular product obsolete. The decline phase may be quick, almost instant – for example, advances in civil engineering technology may render the contractor company's expertise obsolete quickly. In other cases, decline may be slow and steady – as, for example, when client bases gradually look away from their traditional contractors to other service and activity providers.

The product life cycle is also influenced by the following factors:

- **Seasonal:** building and civil engineering work may be retarded or stopped altogether in extremes of cold and wet weather.
- **Locational:** where activities are generated as the result of prestigious, high-profile or regional development work and the public sector pump-priming.
- **Personal:** where, for example, the attachment of a highly valued name or company to a particular project ensures its success; or where work is generated as the result of a high profile or highly prestigious competition (e.g. the Thames Corridor development of the 1990s).
- **Ethical:** the propensity of organisations to set absolute standards of activity whatever the nature of the market in question; and the determination not to compromise standards whether or not this opportunity may be available.
- **General familiarity:** where existing levels of confidence and reputation are used, either to give new products of the organisation their initial launch; or to boost or to regenerate existing products.

The product life cycle is therefore a concept rather than an absolute fact. It is a useful shorthand for both organisational strategists and marketing managers; an indicator of the general state of a given product; a pointer for assessment and analysis that may lead to specific marketing activities being undertaken. It also helps in assessing the product portfolio and the relative merits of each item within it (and this, again, may indicate the necessity or non-necessity for changes and developments and lead to actions being taken).

Product portfolio

The term product portfolio is used to describe a range and mixture of offerings. There are, as always, sectoral norms and considerations to be taken into account. In general, the portfolio will normally include a range of new and existing offerings, well-established and profitable lines, household names, brands and other organisation leaders and flagships. It is also likely to include failures – either items that are becoming obsolete, mature and ageing, or else products that, for whatever reason, have not attracted market confidence and commercial activities. The organisation is also likely to have ideas and products being piloted and tested or undergoing other 'morning star activities'.

The purpose of product mixes and portfolios is to enable the contracting organisation to offer as broad and complete a range of products and services as its expertise and resources will allow and in terms of the demands placed by the client base. Mixes and portfolios should also be seen in the following terms:

- **Product extension:** the mix of products required from the inception of a contract to its completion in order to give the client the maximum possible service and satisfaction.
- **Contractor confidence:** which refers to retaining a broad range of products in order that clients perceive that they are dealing with a sophisticated and diverse organisation.
- **Gap filling:** where the contractor develops new products to fill gaps in the client demand, whether these already exist or are likely to exist as the result of enhanced client expectation.
- **Retention and divestment:** whether it is in fact commercial sense to retain unprofitable products because their divestment would cause more general questions about the capability of the contractor organisation; and whether anyone acquiring the divestment may be able to use this to gain a foothold in the market, thus increasing its competitiveness.

Models of product portfolio analysis are given in Chapter 8.

Branding

The act of branding may be described as striving for differentiated leadership, the creation of a unique distinctive identity alongside the actual product or service:

- *'A brand is a name, term, sign, symbol or design – or a combination of these – intended to identify the products, goods and services of a seller or group of sellers and to differentiate these from those of competitors'* (American Marketing Association).
- *'The essence of a brand is that it is more than an undifferentiated commodity or product. In the perception of buyers, the brand has a unique identity'*. (Randall, 1992).

Brands have the following qualities:

- **The promise:** consistently delivering products and associated features, benefits, quality and services to clients and customers.
- **Attributes:** in the best brands these are always positive – they reflect excellence in the particular sector and include design, materials usage, engineering, furbishment, fitting out; and the quality of the finished product in its environment.
- **Benefits:** branded benefits lead to extreme feelings of satisfaction on the part of customers, consumers and clients; they are a summary of high levels of quality, value, product delivery and service.
- **Values:** this represents the client's ability to deal with a top quality provider; often to purchase the contractor's expertise at high and premium prices; and this, again, is reflected in the satisfaction with ownership and use of the finished product or project.
- **Identity:** using a branded contractor gives the client a clear and distinctive identity, e.g. 'We are using British expertise'; 'We are using a Foster design'; 'We are using Everest double glazing'; 'We are using JCB equipment'.
- **User perception:** this is the feeling suggested to the consumer by associating with a strong brand; e.g. 'We are the sort of client that uses Foster'; 'We are the sort of client that uses Laing'; 'We are the sort of client that uses Caterpillar equipment'.

The following types of brands may be distinguished:

- **Company:** where the company is itself a brand – for example, JCB or Caterpillar.
- **House:** where the product range of a company is branded – for example, HBG Geotechnical Services or Foster Commercial Designs.
- **Individual brands:** this is where the owning company's name often does not appear in the branding process. Offerings are allowed to stand on their own: for example, Butterley Brick is owned by the Hanson Group; ECC, the mining company, is owned by Tarmac; in each of these cases the owning company has no prominence.
- **Generic brands:** these exist as a form of umbrella for a range of products. For example, Everest offer double glazing, and domestic and commercial refurbishment products including doors, windows, office and domestic furnishings made from glass, plastic, metal, aluminium, hardwood – all under the Everest brand.
- **Distribution outlets:** this is the branding of retail and wholesale establishments such as Jewson, Wickes and Redland Brick. It also includes the use of transport and delivery fleets, e.g. Celcon and Nickolls. The point here is that the distributer brand must be at least complementary to the producer brand – if both are positive, the end result is the creation of higher perceptions of quality; where one or the other is not of good quality, the total is adversely affected.

- **Universal brands:** these include Caterpiller and JCB; they are recognised the world over as having a distinctive identity and perception.
- **National brands:** in the UK, these include Laing, Mowlem, Everest and Foster. They have a high level of national identity and perception. More locally, regional and local equivalents of this exist – for example, Pluckley Brick is a very strong regional brand in south-eastern England; Pilkington Glass is a very strong regional brand in north-west England.
- **Specialist brands:** these have no meaning or identity outside their own sector but are very strong within it. The properties of distinctive branding therefore apply to the manufacture of components and parts for assembly, and to industrial business, just as they do to consumer orientation and distribution. For example, the manufacturers of steel stock civil engineering, locks and handles for doors and windows have to use components of a particular distinctive and branded strength often supported by UK, European and global marks of quality.

Strong brands constitute a summary of the quality, value and benefits to be achieved by using that product, service or company. Brand development is a combination of long-term investment in advertising, promotion and presentation supported by top quality product and service delivery. In the architecture, design, planning, building and civil engineering sectors, this is much longer term than for double glazing, domestic and commercial fixtures and fittings, and building products generally, because of the nature of opportunities. Opportunities for brand development are therefore limited and have to be maximised, and this means recognising and using personal and professional contacts, and presentations such as liaison, sales pitches, prequalification, bidding and tendering, and trade fairs as a part of this.

The result of successful branding is as follows:

- a high degree of identity among the target market or segment;
- a strong summary of everything that the company or product provides;
- confidence, reassurance and security – a guarantee of the benefits provided from doing business with the organisation and/or using its products and services;
- future certainty – the organisation or product is permanent and not transient, and will be around 'forever';
- added value – the presence of intrinsic benefits, e.g. of association and identity; and (with components and supplies) the perceived added value that accrues as the result of using these specific items;
- sustainable long-term commercial advantage – the top brands in all sections of industry are able –
  (a) to charge high, often premium prices, for their products; and

(b) to vary their prices and charges according to the nature of the sector being served.

Successful branding is achieved through attention to the following:

- **Quality:** from the client's point of view. Part of this is therefore the reality of quality and part is quality perception. It reinforces the point that expertise, product use and value, and project delivery are not ends in themselves but have to be presented in ways expected and anticipated by the client.
- **Differentiation:** specific attention to those factors that –
  (a) sustain the continuous competitive advantage of the contractor;
  (b) persuade the client to take advantage of these, rather than using the competition;
  (c) identify unique features valuable to the client group or target market which are provided by the brand.
- **Consistency:** consistency of presentation and delivery; consistency of attention to client demands; and consistency of brand identity. Care has to be taken when considering changing the points of consistency to ensure that these develop the brand positively in terms of its identity and uniqueness, and that they enhance rather than dilute customer expectations and satisfaction (see Summary Box 5.9).

SUMMARY BOX 5.9 **Brand Consistency: Examples**

Double glazing
A double glazing company that always fits its windows within four weeks of agreement may not be successful when 'improving' its service to fit the windows within one week, because there may be a perceived loss of product quality, exclusivity and personal attention.

Planning
A planning consultancy may diversify from traffic management schemes into leisure facilities. From the organisational point of view, it is broadening its range of expertise. However, from the client point of view, it may be perceived to be losing interest in its area of considered strength and expertise, and the client may therefore cease to have full confidence.

Architecture
An exclusive top-of-the-range architect or design consultancy trying to break into new areas may choose to acquire subsidiaries for the purpose, rather than risk diluting and repositioning itself through its own name. Clients that have used the practice because of its exclusivity may no longer wish to be associated with something that is known to, or perceived to, have diversified.

- **Continuous improvement:** the problem here is to reconcile market and product/service development with quality and consistency. Whatever is done therefore must build on the existing, rather than repositioning. If repositioning is the intention then either the existing brand strength will be diluted or an alternative approach must be found (as with the acquisition of a subsidiary indicated in Summary Box 5.9).
- **Support:** for building products, and domestic and commercial refurbishment fixtures, this means investment in consistent and high-quality advertising, presentation and promotion. For architecture, building, planning, civil engineering and consultancy this means investment in personal and professional contact, clear documentation, specific customer attention, after-sales and product service, and market research and development.
- **Badging:** the badge, logo, letterhead or site-board is the initial visual point of contact and identity with the brand. Attention is therefore paid to developing this so that an instant summary and highly positive perception are received by the client.
- **Convenience:** this is a combination of accessibility and quality of response. Product and service convenience is always a matter of perception and degree. This is always subject to improvement and refinement. Organisations gain a good part of their strength by making themselves as convenient and as accessible as possible to their customers and clients. The converse of this is that they can lose work very quickly to a competitor that has made itself into 'the most convenient or accessible in the market'.
- **Prestige:** clients like to be associated with prestigious companies and practices. Again, they will gravitate towards these only so long as the reality and perception of prestige are maintained. This is overcome by acquiring or developing a subsidiary when activities certain to lead to the dilution of prestige are envisaged.
- **Client group affluence:** the advantages of branding accumulate according to the relative prosperity levels of the client group and target markets. Where client groups are short of resources, price is the main consideration. Where this is not a problem, clients have a much greater propensity to pay for the convenience, appearance, dependability, quality assurance and prestige of branded projects, products and services (see Summary Box 5.10).

New product development

In reality, new product development goes on all the time, gradually and incrementally. The reasons for getting involved in more radical transformations and for taking a positive commitment to it are:

SUMMARY BOX 5.10 **Advantages and Costs of Brands**

The potential for successful branding may be illustrated as follows:

- brands with 40% market share generate three times the return on investment of those with a share of only 10%;
- for UK grocery brands, the number one brand generates over six times the return on sales of the number two brand, while the number three and four brands are unprofitable;
- for US consumer goods, the number one brand earned a 20% return, the number two earned around 5% and the rest lost money;
- all brands can be profitable: a strong brand in a niche market earns a higher return than a strong brand in a big market; in large markets, competitive threats and retailing pressures can hold back profits of a brand;
- premium brands earn 20% more than discount brands;
- it can cost six time as much to win new customers as to retain current users;
- the best feasible strategy to achieve profitability and growth is to focus on brand differentiation rather than cost and price – although the best strategy in theory is both low cost and high differentiation, in practice it is worth paying some cost penalty to achieve strong branding and differentiation.

Source: G. Randall – *Principles of Marketing,* Routledge, London, 1992

- Decline and obsolescence of existing lines and areas of activity. Ideally, new product development should be in hand before this becomes apparent; it must be put in hand when it is apparent.
- As the result of product portfolio analysis – from whatever standpoint – likely spurs to new product development are: decline of today's breadwinners and the lack of tomorrow's breadwinners; decline of 'cash-cows', too many question marks and 'dogs'; loss of sales of those that make money; divestment of question marks and 'dogs', which the acquirer has subsequently made into winners.
- Market decline, which may be as the result of the market coming to the end of its commercial viability; or of the client base having decided to look elsewhere for satisfaction.
- As the result of product and service transformation by a new or existing player who, in turn, transforms the expectations and levels of satisfaction of a client base.

Contractor organisations must also attend to the following which are more general pressures to produce new products and new ways of doing things:

- The client base demands change, either because of its own first-hand knowledge or else because it knows, perceives or understands that additional benefits are available elsewhere in the world sector or industry.

- The market leader has produced a new, improved or (very occasionally) revolutionary way of doing things that requires the rest of the industry to follow it.
- New technology is available, which means that existing and equivalent products can now be produced in more cost effective, cost efficient ways.
- The distinctive position of a particular organisation is that of pioneer and innovator, and there is a consequent commitment to maintain this position.
- Specific demands from clients for project and product longevity, durability, quality and size often lead to the derived demand for innovation.
- The use of both in-house and other research facilities (e.g. universities, databases, involvement in other pioneering and pilot studies) to pursue ideas to see where they lead; the commitment here is the pursuit of knowledge, understanding and enlightenment, and the pursuit of commercial advantage is almost a by-product.
- The knowledge that a peripheral activity is worthy of developing into a full commercial offering.
- Client-based demands of an enhanced total service from its contractors; and some of the components of this may have to be acquired or developed very quickly – this is different from peripheral activity in that it almost invariably leads to an enhanced level of commitment, cost and resources being required.

New product development takes the following forms:

- **New inventions:** e.g. computer-aided design and architecture; RIC; polystyrene/concrete cladding.
- **New technology applied to existing products:** for example, databases rather than paper files and archives.
- **Speed and accuracy of response:** again, often through electronic capability, the ability to process requests and large volumes of data very quickly.
- **Improved technology:** resulting in enhanced quality, volume and flexibility of output, better resource utilisation and improved cost, efficiency and effectiveness.
- **New designs and new facilities:** for existing and hitherto accepted ways of doing things.
- **Products and services:** new to this industry and its markets but which exist elsewhere; new to this organisation but which exist elsewhere in other organisations in the sector.
- **Repackaging:** re-presenting existing products and activities (in the same or to new markets).

- **Adoption of an existing range of activities:** developing the expertise to market and deliver them effectively within the particular client base.

Successful organisations take their own distinctive view of new product development. It is likely to reflect a part of their overall marketing position as follows:

- **First in the field:** taking the distinctive view that the advantage in pioneering new products, projects and services with all their inevitable teething troubles outweighs the fact that followers are likely to learn from these mistakes.
- **Second in the field:** and whether to be a close or more distant second – in which the distinctive position taken is to have a new product ready to launch provided that the pioneer is successful – and the capacity to withdraw it if the pioneer is not successful.
- **'Me too' and 'all-comers':** where the new product will be commercialised only when it is widely accepted in the field.

Successful new product development

The first step to understanding successful new product development lies in pinpointing the main reasons for failure. There are general constraints that have to be recognised as follows:

- **Short-termism:** organisations that want instant results are unlikely to commit themselves to the long-term resource, technology, expertise and finance demands necessary to get successful new products to market.
- **Horns effect:** the distinctive feature here is the 'blip' (or series of 'blips') which often come to be seen as product defects rather than teething troubles. The effect is usually to cause loss of confidence and, therefore, the new product gets hammered.
- **Top management:** top management kills new products, either because it does not understand them; because it is remote from the day-to-day activities; because it has an historic (rather than current and/or future view) of the industry; or because it sees no reason to change (and there may or may not be reasons to change).
- **Accounting practices:** new products do not fit easily into accounting schedules, nor into accounts-driven budget constraints. The problems arise when these schedules and constraints decree that the new product has to show positive results by a certain date or within a given budget; and that where this has not occurred, it is hammered on the orders of accountants rather than production or marketing experts.
- **Time to market:** this is the difference between customer and client demand which is often instant, immediate and short-term, and the

ability to satisfy that demand which may not be quite so instant. In many cases, and in all kinds of organisations within industries, getting the time, resources, technology and expertise – and the go-ahead – to develop products for which demand is proven, well known or strongly indicated, often requires a marketing campaign all of its own.

- ***Realpolitik:*** problems of getting resources and new product approval are exacerbated when the particular form of organisation politics requires that new product developers have to lobby for the ear and support of those with influence, those who control resources and those who can block progress. These problems occur basically for two reasons: because top management is not aware of them and is therefore unaware of the problems; or because top management is only too aware of them because it uses this approach as its chosen means of 'managing' the organisation.
- **Frustration:** in relation to new product development, frustration occurs at two levels –
  - the product champions, those who have faith in what they are doing and of the integrity, value and benefits of the innovation once it is brought to market;
  - the marketing and sales staff who have spent time and resources researching, piloting and promoting the innovations, only to have them repeatedly delayed or cancelled.

  The medium to long-term result of this is to drive the creativity, capacity for innovation, drive and enthusiasm for this area of activity from the organisation. It is also invariably symptomatic of a wider lack of commitment to any form of innovation or progress.

The correct culture, climate and environment have therefore to be created if any form of new product development is to be successful. This is because the presence of any of these constraints is certain to dilute its effectiveness. The more of these that are present, the greater the inability to innovate effectively (see Figure 5.4).

More specific reasons for new product failure can then be pinned down. These are:

- Lack of genuine product/service uniqueness; inability to deliver required benefits; benefits delivered are not those required by the client base.
- Mistaking general positive responses for clear commitments to buy or use – the difference between would you and will you; a true level of consumption, and therefore income, has been assessed in general and not specific terms.
- Product performance – that the product has no distinctive advantage once it comes to market; that it is unable in practice to compete on technical aspects, value, benefits, quality, cost or price; that while it

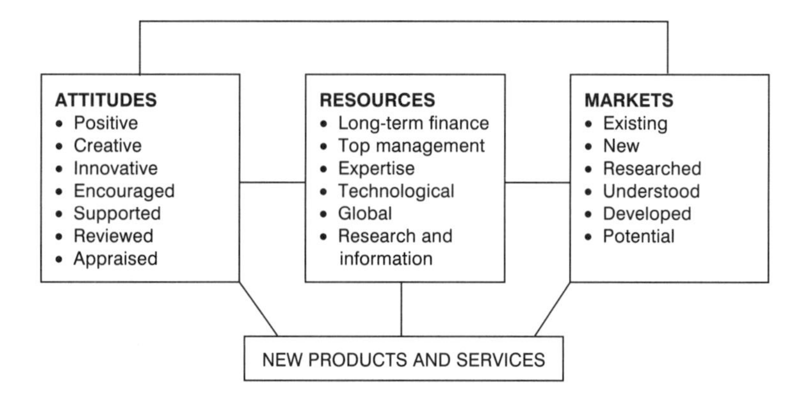

Figure 5.4 Creating the conditions for effective new product development and innovation.

is no worse than the competition, the competition is at least well established; or conversely, that the new product receives instant high levels of recognition and usage which the developing organisation is unable to satisfy and this, in turn, leads to dissatisfaction with the organisation's service levels and opportunities for others to mop up the spare market capacity.

- Competition performance – that competitors either get their own version of the new product to market more quickly, more cheaply, in greater volume and in better quality; or that by anticipating changes in the given market, the competitors have come up with a superior and distinctive offering.
- Timing – this normally means that the innovation has taken too long to come to market and this, in turn, has led either to a competitor being able to satisfy demand, or to the client base finding alternative means of satisfaction.
- Lack of commitment to the process of new product development is normally coupled with lack of resources, staff, technology, expertise and championing.
- Poor planning and research, misperception and misunderstandings of the needs and demands of the client base; working on perceptions and received wisdom rather than on hard information.

On new product development, Tom Peters made the following remark: *'If you were a normal thinking, reasoning person concerned with new product development, you would never start. Because statistically the chances of getting something from inception to market are zero.'* (Tom Peters – *The World Turned Upside Down*: Channel 4, 1986).

The point is, therefore, to create the conditions and make the commitment necessary to give new product development the greatest possible chance of success. Attention has then to be paid to the following:

- the right priority in the broader strategy and marketing management for new product development;
- the right overall approach to innovation and development;
- high-quality information and research leading to correct identification of customer needs and wants, and to the value, volume and nature of customers to be served;
- long-term commitment;
- adequate cost and budget management;
- diversity, multiplicity, speed and flexibility of approach;
- review at every stage, coupled with speedy response – positive and negative – from those in authority.

Companies with successful records in new product development review their innovations along the following lines:

- the extent to which the new product is unique or superior, either in performance or perception (or both);
- that they have fully understood the nature of the market – and have targeted their marketing directly at this;
- that a new product is complementary to the existing range of products; *or* that a distinctive identity (often in the form of acquisition or creation of a subsidiary) has been undertaken where desirable;
- that a new product does not dilute the satisfaction of existing customers; that the new product does not affect the relationship with satisfied customers;
- that the new product is targeted either at a new need of an existing customer or at existing needs of new customers;
- that the nature of the marketing requirement is understood in advance – and that resources and energy are committed to it.

It is to be emphasised that this is a means of review and evaluation only – these are not ground rules for the successful implementation of new products. However, if this form of review is implemented at the time of developing the product, piloting it and getting it to market, it minimises the chances of failure; and any form of review and evaluation maximises the chances of success.

New product development is also only successful if the internal capacity shown in Figure 5.5 is available.

New product screening

The screening of new products is undertaken to assess organisational capacity to produce some sufficient quality and volume at costs that

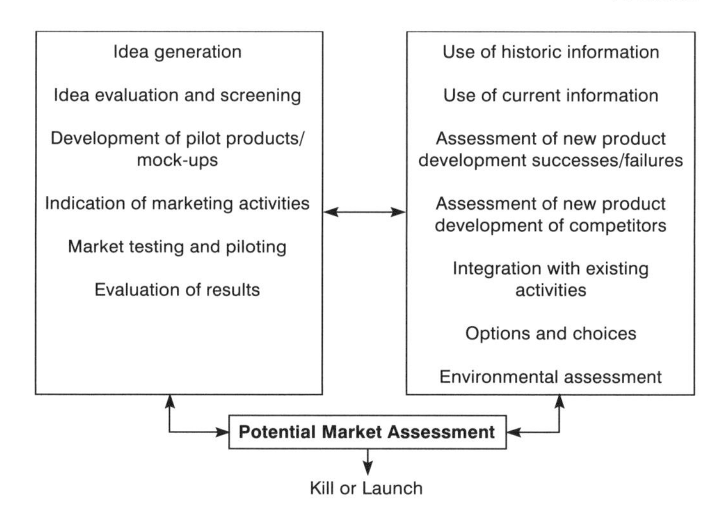

Figure 5.5 New product development: internal capacity.

make them commercially viable; and to identify market segments large enough and with the capacity and willingness to buy them. The screening process is not exact and is often carried out (with varying degrees of effectiveness) on incomplete information and general and perceived market understanding. This is especially true of genuine innovations where the particular contractor may only have its own current reputation for success on which to base its faith in the new product, since the new product itself has no track record. The use, application, mix and priority of criteria clearly vary between organisations and different sectors in the industry. In general, they may be seen as follows:

- **Technical and technological feasibility:** the capability to produce to the required volume and quality as above.
- **Motivation:** the willingness to produce to the required volume and quality as above.
- **Current market size and capacity:** the ability and willingness of the client base to spend on the new product in the immediate and medium-term future.
- **Potential market size and capacity:** the likely duration of the market; the likelihood of expansion or contraction; susceptibility to inflation and destabilisation; the effect of political and macro-economic actions.
- **The threat of entry:** of greatest concern to pioneers and early players in the field when considering 'me too' and 'all-comers', and the likely and possible effects of this; or the threat of entry from international and multinational players which may destabilise the whole sector.

- **Entry barriers:** identifying these and the level of commitment and investment necessary to overcome them.
- **Exit barriers:** identifying the costs and consequences of withdrawal if it becomes necessary.
- **Returns:** the best, medium and worst returns achievable; the lowest level of return that the contractor is prepared to stand in return for becoming involved; the nature and levels of investment necessary to achieve this; payback periods and the reasons that these have been (or are to be) established.

The result of this is an informed, managerial and marketing discussion and evaluation, the conclusion of which is either:

- to continue with the new product and get it to market;
- to discontinue the new product and put the knowledge, experience and understanding gained to use in the future;
- to conclude that the new product in its current form is not right and to continue to bring it up to commercial standard.

Other factors in new product development

These are as follows:

- **Market testing:** trying the new product out on a small niche of the market to test it for initial responses.
- **Mock-ups:** to have a version of the finished product available for inspection by potential clients.
- **Piloting:** this usually means producing a small volume of a commercial standard product and testing it on specific subsectors of the market or client base; assessing both product performance and client reception; using this information as part of the basis from which the new product is either developed further, commercialised fully or hammered.
- **Speed to market:** any specific pressures, especially those relating to competition and the capacity of competitors to dominate the market if the product is not launched quickly; or conversely, that jumping the gun may lead to an imperfect product, inadequately launched and marketed.
- **Availability of products:** this closely relates to the speed to market, but is concerned with the volumes that can be made available within particular time frames. This refers first to the short term and the ability to satisfy immediate demand. Attention has then to be paid to the longer term and the ability to satisfy sustained demand and also to provide replacement parts, maintenance, upgrading, refurbishment and other aspects of product support and after-sales service.
- **Further research:** the real or perceived need for this; what else it

may reveal; who has commissioned it and their agenda (supportive, destructive or hedging their bets); what the further research would cost; and the length of time it would take.

- **New product performance presentation:** the initial and continued ability to present the new product in terms of the benefits required by the client base; the recognition of the extrinsic quality and merits of the product itself, these have still to be commercialised – marketed, promoted, packaged and presented – effectively to the particular market.
- **Other factors:** chief among these is the safety of the product and the certainty and reliability of its performance; the absolute quality and expertise of service; the extent of risk, of product or service failure or lack of adequacy; the requirement for product quality validation (e.g. BS and ISO marks); and any distinctive or specific legal checks necessary (see Summary Box 5.11).

SUMMARY BOX 5.11 Key Factors in New Product Development

All other things being equal, the key factors are the predictability of the state of the market in the medium to long-term; and the time that a new product takes to get to market.

State of the market

The main constraint for the construction industry here is the inability to control so much of its operating environment. The construction industry environment is heavily influenced by the acts of politicians, interest rate fluctuations and the real or perceived ability to spend large amounts of money for long-term results. This is also a key feature of the confidence of both commercial and domestic building and house purchasers – their perception or willingness to afford this level of commitment over the long-term.

Time to market

As changes in commercial tendering processes occur, the capability of organisations to respond effectively is called into question. Part of this relates to the construction industry's traditional view of itself as being product and expertise led, and the consequent need to transform its attitudes. Part also is concerned with the processes that particular contractor companies have either grown or allowed to have grown, which almost invariably get in the way of speed of response. The fundamental issue is that the construction industry's clients, as with the customers of all sectors, are having their expectations and levels of satisfaction constantly raised and a key feature of this is the speed of response.

Beyond this, the key to effective and successful new product development lies in having the right and positive attitudes that come from adapting marketing as the key to effective and profitable performance. Stripped of perceptual constraints that have arisen as the result of traditional ways of working, there is no reason why the UK construction industry cannot become far more flexible, responsive and dynamic in its dealings with both its domestic client base and also in seeking opportunities overseas. Above all, the best Japanese, American, German and French building and civil engineering companies have gained market share in the UK and elsewhere at the expense of the domestic industry. This is not because the domestic client base wants to use foreign competitors. It is because the external product quality is equally as good and the product delivery and support service are vastly superior and more innovative – and generally much quicker, more flexible and more responsive.

SERVICES MARKETING

For architecture, design, planning, project management, consultancy and agency activities, the ability to market a concept of service has always been critical. For the rest of the construction industry – civil engineering, building and building products – this is becoming evermore critical as client bases demand ever-increasing levels of product and project support as an integral part of the contract.

From a marketing point of view the service is: any act or performance offered by one party to another that is essentially intangible and does not result in the ownership of anything; though service may be related to a physical product (Kotler, 1993).

For the construction industry, the following types of service provision may be identified:

- **Tangible products supported by services:** in which the services offered enhance client appeal for the product. This relates to all aspects of building, and civil engineering where the creation of the particular facility is supported by project and environment management, client liaison and specialist consultancy services; and sometimes also through maintenance, refurbishment, upgrading and facilities management services; and for public projects, sometimes also more general services to the wider community (e.g. education and sponsorship).
- **Product–service hybrid:** where the product and service are offered together. This occurs, for example, in the domestic sectors where double glazing, kitchen, bathroom, bedroom and other fixtures and fittings providers offer a generic range of products specifically tailored and fitted on a unique and individual basis.

- **A service with tangible product elements:** this occurs, for example, in the building and civil engineering sectors where regularity and frequency of delivery are of greater importance than the product itself, e.g. where there is a shortage of storage space. The point is that if the means and frequency of delivery – the service – could not be guaranteed, the supplier would not have been engaged.
- **'Reported service':** in which the service is intangible but supported by documentation that gives substance to the contractor–client relationship and provides points of reference for the service quality and operational relationship. This is the nature of service provided by architecture, design, planning and problem-solving consultancies.
- **Pure service:** in which the service is completely intangible. Examples from elsewhere include the services of medical practitioners and social security. In commercial situations 'pure service' is extremely rare, confined to urgent problem solving, e.g. a service provided by an arbitrator after a time of site industrial dispute in which the arbitrator uses its expertise to get the problem resolved. While this may be the extent of the arbitrator's involvement, it would in practice be very unusual for the matter not to be reported and written up subsequently.

Some key points need to be addressed:

- the nature of the service demanded by the client; the nature of the service provided by the contractor; and the similarities and differences between the two;
- the nature of the relationship necessary for effective and profitable business between contractor and client;
- the extent to which the service can be offered in generic terms, and the extent to which it has to be customised either to groups of clients or individual clients or for individual jobs;
- the nature and balance of supply and demand for the services;
- how, when and where the service is delivered; the key features of effective service delivery; the key perceptions of effective service delivery;
- contractor and client pressures and priorities, and the reconciliation of these.

When viewed in this way, it becomes clear that everyone in the industry needs to be aware of the need to provide effective and quality services, either in support of tangible products or as the main part of the offering – and even for building and civil engineering companies where product quality, durability and deliverability are more or less assured, whoever is commissioned, the ability to gain the work may indeed hinge on the levels of service that the contractor is prepared to provide.

From a marketing point of view, services have the following properties:

- **Intangibility:** services cannot be seen, tasted, felt, heard or smelt before they are bought. Clients therefore look for other signs and evidence of quality and value of the particular service – and these include reputation, track record and prices charged. Cues are also taken from the nature and quality of any initial consultation, the nature of the service provider's hospitality, offices, technology, equipment and other factors.
- **Inseparability:** services are normally provided and consumed at the same time. The service provider has the capability to offer the service all the time; but the only time it is delivered is when and where the client requests it.

  Some services delivery is in the client's presence, e.g. the presence of a consultancy report, planning or architectural proposal. Others are not, e.g. a client in London who owns a factory in Manchester for which there is a facilities management contract with the contractor, simply contacts the contractor when there is a problem and leaves the solution to the contractor.
- **Variability:** on the one hand, service offerings are generic – design, architecture, quantity surveying, planning, consultancy, after-sales, facilities management are all clear definitions of activities within the industry. On the other hand, the nature, level, volumes and quality of service offered vary widely – and are also perceived to do so by client bases. So the problem lies in determining and standardising a high quality of service that can be made universally available upon request. This is achieved through investing in high-quality staff; continued attention to the nature and levels of client demand; and allocating teams and groups of people to particular specialisms, clients and locations. The end result of this is that enduring, combined, personal and professional relationship can be developed. This also means that problems of variability are anticipated and dealt with in advance of any marketing activity – and responsibility for this rests firmly with the contractor.
- **Totality:** this refers to the totality of the service on offer; and the totality of the service demanded by the client. It is the extent to which the equivalent of 'one stop shopping' is offered. This is represented by a combination of the nature of the expertise, support services also provided, and the capability and willingness to anticipate and tailor these – and where necessary, or desirable, or requested, extend them – to the demands of particular clients. This, in turn, involves knowing and understanding the normal, likely and possible service needs of the client base; and it means taking active steps to develop the fullest possible service provision in support of the particular expertise. Many clients prefer the idea of 'one stop shopping' and will often gravitate towards companies and practices that pro-

vide this, rather than having to engage a wider range of service organisations.

- **Perishability:** services cannot be stored or stockpiled. They must be made available when clients demand and if clients cannot, or will not, wait for the service provider, they will go elsewhere and the opportunity is lost. For those organisations engaged in service provision, therefore, the presence of top quality staff and equipment technology necessary to provide the level of service is a fixed cost – one that must be borne whether or not work is actually being conducted. Moreover, from a management point of view, steps have to be taken to minimise the problems of perishability and these include: guaranteed minimum response times that give a certain amount of latitude if and when staff need to be reorganised; premium prices in return for instant responses (part of which may be used to hire in part-timers and subcontractors to cover peaks in demand); differential pricing, encouraging the client base to take a long-term view of the services demanded (e.g. if a design is required in one month, it costs a quarter of the price charged for the same design in one week); organisations can have a pool of part-timers and subcontractors available to cover peak periods to ensure continued universality and quality of coverage; task and client prioritising so that key employees know absolutely where and how they are to be deployed during periods of pressure.

This has then to be related to the following:

- **Access:** a form of access – whether it is through personal call, telephone, fax, e-mail or a combination of these – that is known, understood and convenient to both contractors and clients. To this there may also be added –
  - time constraints – such as guaranteed response times and other deadlines;
  - the quality of the initial contact – whether this is always direct to an expert, or to someone who has the useful working knowledge, or to a receptionist who in turn guarantees to hand the problem on and get a response within a certain time;
  - variability of response – this can be designed into the service and includes the use of helplines to handle general enquiries; reference lines to gain access to specific expertise; and call-out facilities for immediate problems, crises and emergencies.
- **People:** some clients prefer to have the same personal point of contact and service provider in all their dealings with the contractor, and this should be the aim wherever possible. Where it is not possible, the level of expertise must be maintained as failure to do this always leads to the client casting around for alternative sources.

- **Technology:** the highest levels of service are normally delivered by expert staff using the best possible equipment for the work in hand. This enhances speed as well as quality and value of the particular service.
- **Literature:** this has to be of a quality sufficient to reinforce the service provider's desired perceptions in the mind of the clients. A part of this normally has to be customerised – devised and presented to specific clients and tailored to their particular needs.
- **Continuous improvement:** this has reference both to the need constantly to improve and update expertise, technology and quality of service provision; and the need for attention also to more general service features, such as speed and quality of response, anticipating the client's needs and extending the depth and breadth of the service wherever possible.

Service expectations

Service expectations are best understood from the client's point of view. Clients retain service providers because they expect to have problems addressed and solved that their in-house expertise does not cover. The view is taken that it is better to pay high – often very high – prices as and when the service is required, rather than retaining the expertise full-time in-house.

Because of the nature of services, satisfaction is apparent only at the point of completion. Between initial commission and service end, a range of gaps and problem areas may be identified, as shown in Figure 5.6.

It is clear from this figure that there is great potential for misunderstanding and misperception, and this may lead, in turn, to misdelivery and dissatisfaction, lack of thought or incomplete satisfaction.

To give the greatest possible opportunity of avoiding this, preparation is everything. The effective marketing of services must build in the greatest possible time for preparation, briefing, discussion and analysis. This should also apply to crisis management and troubleshooting activities, where a key quality of the troubleshooter is the ability to master a brief and get at the core of the problem or key information very quickly.

Having established an accurate and well-informed brief, clients expect the following from service providers:

- **Reliability:** the ability to deliver that which was promised dependably and accurately.
- **Responsiveness:** the ability to respond quickly, flexibly and dynamically to briefs and commissions.
- **Confidence and trust:** in the expertise and application of the staff sent by the service provider.
- **Empathy:** working at all times with the client's needs uppermost. The best service providers never make excuses to clients.

1. EXPECTATIONS GAPS

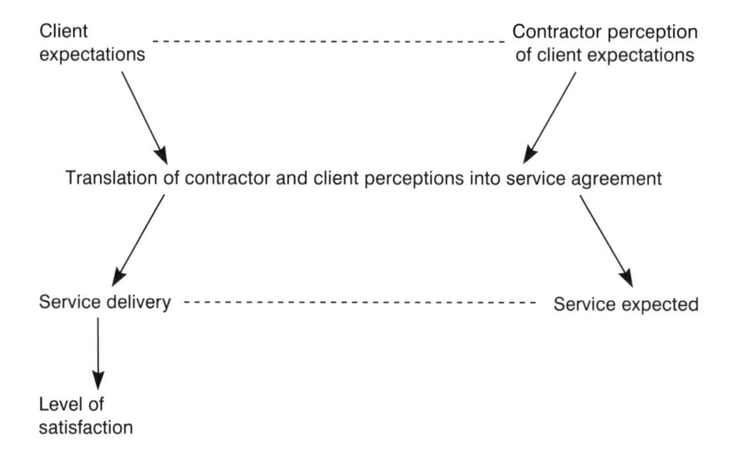

2. PERCEPTUAL GAPS

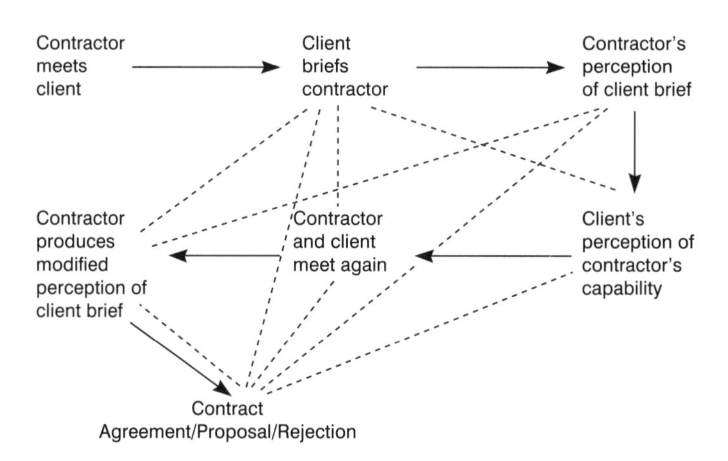

If each arrow represents a gap, then at least six gaps – potentials for error and misunderstanding – exist.
Each broken line represents the capability to refer back to previous discussions – which may either clarify issues, or cloud them still further.

Figure 5.6 Service gaps.

- **Tangibles and visibles:** the conduct and demeanour of the contractor's staff when working on the client's premises and facilities.
- **Focus:** an obsession with customer service in the contracting organisation. This is established in the attitudes of top management and is all-pervasive throughout all staff.

- **Remedy:** any organisational problems that arise during the service agreement are put right during the course of the work.

Effective service provision

The key to being an effective service provider lies in having high-quality, expert, flexible and responsive staff. A very high proportion of the staff will be working directly with clients on problems. All staff need the attitude of placing the customer at the centre of the scheme of things.

Front-line staff are supported by high-quality, flexible and responsive support staff, and by the best technology and equipment available.

To be an expert service provider therefore requires specific points of attention and investment and this is essential for continued and future success.

Components of customer service

These may be summarised and itemised as follows:

- response times;
- response quality;
- time from order to delivery;
- reliability of delivery;
- service availability and continuity;
- contract completion;
- quality of support staff;
- convenience of placing order;
- accuracy of receiving order;
- quality of salesforce;
- attitude of salesforce;
- call and contact patterns of salesforce;
- payment and invoicing methods;
- quality of presentation;
- consultation and liaison during service period;
- consultation and liaison on service improvement;
- service range reviews;
- co-ordination between service provider staff and support staff;
- variety and convenience of points of access.

6 Place, Location and Outlet

In marketing, *place* represents the meeting point between the buyers and sellers of goods and services. At issue is the need to ensure that what is offered is as accessible and convenient as possible to customers and consumers, and more accessible than the offerings of competitors.

In the marketing of construction industry activities, place has a range of connotations – from the use of established retail and wholesale outlets for building materials; the creation of personal and professional contacts as the meeting point in tendering and contracting processes; to the creation of an almost psychological place in the marketing of design, planning and consultancy services.

In the marketing of construction goods, therefore, place is a clear concept involving choosing preferred outlets and means of distribution to those outlets from a universally familiar range – shops, mail order, warehouses, catalogues; increasingly the use of journals, TV, radio and computer-based facilities.

In the marketing of primary construction work, related disciplines and activities and also full consultancy services, the basic premise – of choosing preferred outlets and the means of distribution to those outlets – remains the same. Their operation is fundamentally different however. The key features are access and convenience.

ACCESS

For consumer goods, access is largely a matter of degree. If one channel proves ineffective, it is possible to drop it and change to others fairly quickly. Where it is successful, the relationship between supplier and distributer can be developed quickly to mutual advantage, for example through the use of discounts, pricing advantages and other hook-ups that relate to the rest of the distributer's offerings. It is also possible for the supplier to prevail upon the distributer to give priority positioning and emphasis to its range of goods above the rest of those on offer at the distribution centre.

This is possible to an extent at the retail end of the construction industry – double glazing, small volume building products, operations of builders' merchants – and these are now discussed in detail. At the capital service and expertise end of the industry, the key to successful access lies in the creation of a high-quality, continuous and professional relationship between the buyers and the sellers of the products. Three levels of this relationship may be distinguished:

- **General familiarity:** based on promotion, literature, public relations activities, features, articles and inserts in the trade press.
- **General confidence:** based on a good, sound and positive attitude held by the client base towards the contractor; a reflection of the 'general positive attitude' referred to elsewhere.
- **Direct familiarity:** based on direct sales efforts and targeted promotion, literature and public relations activities.

In each of these cases the point of access achieved is roughly the same as that achieved by Coca Cola for its drinks products. This is based on wide general knowledge, general ease of access – and absolutely no pressure to buy or sell. The basic difference lies:

(a) in the nature of the product (where Coca Cola is a low cost, low risk, unconsidered purchase – construction is the opposite to this; at the commercial end, and even at the personal retail aspect, is a much more considered purchase with much greater consequences);
(b) in the actual point of access, which (except for the operations of builders' merchants) is an office, council chamber, political forum or potential site; and for domestic refurbishment and upgradings such as bathrooms, kitchens and double glazing, the consumer's home or contractor's showroom; and for commercial refurbishment, upgrading and transformation, the facilities in question and again the contractor's showroom.

The necessity is therefore, to identify a range of points of access as follows:

- **Initial entry:** through the public relations contractor liaison functions of the client or potential client and which starts to build on the twin requirements of general familiarity and direct confidence.
- **Product and service development:** through being passed on by the PR and contractor liaison functions to those who screen potential contractors; this is the first step towards developing specific points of access and starts to build direct familiarity.

It is then a question of being handed on to those who commission work – contract managers, project instigators, financial directors and managers, end-user interests, vested interests and lobbies, and other eco–socio–political groups and interests.

This step is essential in the marketing of high-cost, high-risk, fully considered purchases and constitutes an additional fourth level of the access relationship – direct confidence – which is the key to translating a generally favourable response into profitable work.

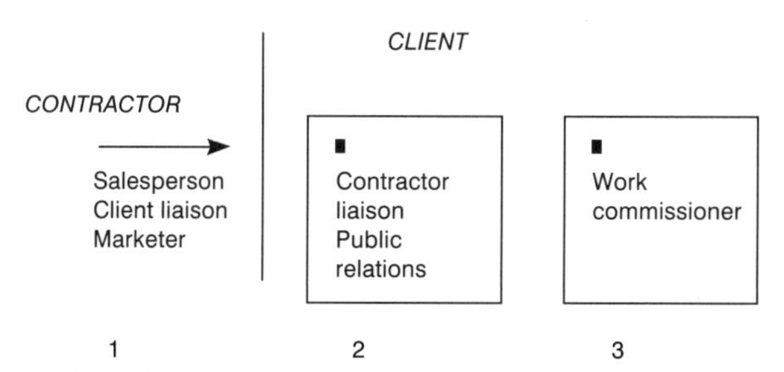

- Relationship between 1 and 2 is easy to achieve, and non-productive
- Relationship between 2 and 3 is unknown
- 1 needs 2 to get to 3
- Critical relationship is between 1 and 3

Figure 6.1 Points of access.

Direct confidence is the outcome of efforts designed to secure a mutually profitable understanding and interdependency; the ability to match contractor expertise with client demand; and supported, in turn, by mutuality of direct interests between contractor and client and with reference to important and influential vested interests.

Figure 6.1 illustrates an overview of the points of access available throughout the whole process in the construction industry.

CONVENIENCE

The second issue is to ensure that whatever place is on offer is as convenient as possible to the prospective client. Again, the problem is to translate the elements of convenience provided by shops and supermarkets for consumer goods into terms practicable for construction industry activities.

Clearly, access is one of these features. The others are time; location; replaceability; and after-sales and guarantees. In turn, these reinforce – and are reinforced by – general and direct levels of familiarity and confidence.

Time

Time frames are normally set by the client. Where this varies, it is ideally by agreement between contractor and client. It is possible in some circumstances for the contractor to impose its own time frames on the client, though this is not normally the precursor to a profitable long-term business relationship (see Summary Box 6.1).

SUMMARY BOX 6.1 **Time, Luxury and Prestige**

There is a perception (not confined to the construction industry) that 'the best things in life are worth waiting for'. It is used by luxury, high-profile and high-prestige companies in all sectors to help to generate and reinforce myths and exclusivity around what is on offer. Without exception, this is only profitably sustainable where long-term steady demand outstrips short and medium-term supply.

Location

The problem here lies in substituting the factors of location afforded by the presence of shops for consumer goods. Clearly, this goes much deeper than mere presence which simply gives availability and choice. Presence, availability and choice have to be covered in such a way as to meet the client's propensity to use. This is based on a combination of familiarity and confidence, indicated previously. This is in turn reinforced by adequate and effective channels of communication (all channels – written, spoken, electronic, general and specific); and supported by quick and direct personal access and face-to-face contact according to the needs, wants and demands of the client. Failure to do this simply means that clients will exercise their capability to choose – and choose elsewhere if that is possible. So while it is not possible to provide universal access by location, it is possible – and essential – to cover the demands for the convenience factors and advantages that arise by having a wide range of outlets available.

Replaceability

For consumer goods this is straightforward – the product is returned either to the point of sale or to the manufacturer for replacement and/or compensation. UK and EU Law also allow consumers to prosecute both outlet and manufacturer for the sale of unsatisfactory, poor-quality goods and services, and for goods and services that are (or are perceived to be) misrepresented in some way, and this includes both the products and services themselves and also the benefits that they are supposed to afford.

In many ways, covering this point is more straightforward for the construction industry. There are specific legal constraints covering a wide variety of the industry's disciplines and activities – for example, design, civil engineering, commercial house building and refurbishment, many of which are underwritten by industry norms and standards and reinforced by professional bodies such as the Federation of Civil Engineering Contractors, the National House Building Council and the Glass and Glazing Federation.

Beyond this, however, and given the constraints of the industry at the end of the 20th Century and beginning of the 21st Century, it is absolutely essential that a responsible attitude is adopted. Not to do so simply gives advantage to all the others in the sector who do. It also simply does not meet clients' expectations or social conventions and, again, this invariably leads to loss of business.

Moreover, any gaps in this part of the contractor–client relationship inevitably become a matter of litigation when things do go wrong. This can – and does – take years to resolve in many cases. When this happens, the reputation of the industry as a whole suffers. It is very susceptible to adverse media coverage; and the loser is always the contractor and sometimes the client also (see Summary Box 6.2).

After-sales and guarantees

The traditional convention of the industry has been not to include after-sales activities. Part of this is culture – the organisation moves on from one project to the next, leaving the previous work behind. Part of this is due to the fact that it was seldom, if ever, demanded. Moreover, in most cases, after-sales actually meant the remedying of faults and teething troubles in facility operation and there was – and remains – a well-established claims procedure supported by litigation where necessary.

Parts of the industry have been transformed by the demands placed on preferred contractors by commercial clients, e.g. design, build and fitting out contracts for supermarkets, shopping centres and retail parks. General perceptions have been affected by the fact that US, German and Japanese contractors include extensive after-sales as an integral part of their quality of service. Local and regional house and public facility builders tend to offer after-sales anyway, so that if things do go wrong they can be put right without critical loss of local reputation. More generally, it is a reflection of what is expected by all customers and clients in all walks of life.

The coverage of the after-sales package often varies between organisations and work, and will normally be based around one or more of the following:

- **Fault remedy:** to a schedule agreed as part of the contract.
- **End-user induction:** so that those who are to use the finished project are fully familiar with it.
- **General maintenance:** both preventing and curing faults.
- **Facilities and environment management:** the acceptance/agreement of a range of responsibilities when the facility is in use; and this may include maintenance, refurbishment, technology and usage upgrades; and more general features such as cleaning and security.

SUMMARY BOX 6.2 Guarantees: Building and Construction on Landfill Sites

Building on landfill sites is attractive to both contractors and clients because the price of land tends to be lower and the consequent opportunity for profit margins much higher.

Serious problems arise either because the site has not been allowed to settle fully, there is the existence of substructures to former buildings or because of the presence of toxins as a legacy of its previous usage. Those working with such sites have therefore to address the problems of subsidence, site clearance and continuing pollution.

Such problems are normally covered by organisational indemnity insurance and the guarantees of professional bodies (e.g. NHBC, FCEC). However, this does not prevent problems from arising, nor does it guarantee that they will be resolved to the satisfaction of all concerned. As examples:

- a medium-quality housing development in north Kent in 1986 had to be scrapped after the foundations of some of the houses started to fracture because the recent landfill could not support their weight;
- an industrial estate in Newcastle-upon-Tyne was completely laid out in 1991 before it was discovered that there existed a residue of toxic and lethal chemicals dating back over 100 years which could not satisfactorily be dealt with because nobody knew who had put them there in the first place;
- part of the problem with redeveloping the rest of the London Docklands is due to the fact that the integrity and quality of the building land cannot be guaranteed – especially those areas that were used for military installations where no records exist of what was deposited on them;
- a housing development near Yeovil in Somerset in 1996 has had to have its foundations fitted with costly gas exhaust pipes and associated works after the refuse tip on which the development was built started to generate methane gas.

In each of these cases the work was underwritten by the various guarantees. However, this does not prevent serious disruption to the end-users and therefore the client's reputation; such cases also generate continuing adverse media coverage which dilutes the quality of relationship and mutual trust that should exist between contractor, client and end-user.

- **Options on future site redevelopment:** if it ever becomes essential or desirable.
- **Staff training for the end-user:** where there are specific, technological or legal requirements.
- **Continuous improvement:** of the facility, including attention to materials, energy, technology management and the quality of working life.

The activities are mixed and emphasised to meet the real and perceived needs of the particular client group. For some activities, it is essential to recognise also what can possibly go wrong so that at least there is an element of fore-warning and therefore fore-arming.

It is also necessary to determine whether after-sales and guarantees are to be built into the contract price or else offered as add-ons should the client want them. How this is done depends on client demand and expectations, and will become apparent long before it becomes an issue as the result of having built a high-quality, effective and professional relationship.

Two other key features that relate to the place aspect of marketing should be noted. These are:

- time to market;
- personal contact and access.

TIME TO MARKET

This is the speed of response of the contractor and it applies to all aspects and activities of the industry. It is less of a problem where a client's time constraints are not a priority. It is a real marketing issue when the client wants the work started and/or completed quickly. There are both general management and sector specific factors.

General management

All activities and procedures are capable of improvement subject to the availability of technology, expertise, creativity and willingness, and the adequate capital base to support them. All other things being equal, the keys are creativity – the ability to use the available resources in ways that will meet this client demand; and willingness – the motivation to do this and accord the client the priority requested.

There are implications for this. The main point is that any organisation that does transform its time to market revolutionises the expectations of its client base. Today's transformation becomes tomorrow's norm. Where one organisation can achieve this but others in the sector cannot, it can lead to sectoral destabilisation. In general, however, the ability to compete on 'time to market' is a distinctive advantage. As well as client satisfaction, it improves contractor cashflow, especially where the bulk of money is not paid until the contract end.

Sector specific factors

There are also sector specific factors as follows:

- **Architecture and design:** speed of effective response leading to initial proposal and/or final and acceptable design, especially organisational attitudes to instant and short-term requests. The attitude to these is balanced against perceptual problems of being 'too quick', i.e. if the response is too quick, it cannot have been properly thought out.
- **Planning, project management and other consultancy and agency services:** the present state of this part of the industry is such that an initial response to a request is normally expected in a few days at most; and the main exception to this rule is physical distance (e.g. a request for a meeting from an organisation in Malaysia to an agency in London has to allow time for travel!). Otherwise any organisation in these sectors that does not respond quickly must recognise the disadvantage at which it is placing itself.
- **Building materials and supplies:** there are two issues for this sector. One is to have sufficient access and distribution support to be able to meet the widest possible anticipated range of demands. The second is to be able to tailor the service to specific clients. For example, just-in-time type deliveries – small, regular and precisely scheduled – are increasingly the norm on major civil engineering projects. This sector is also being transformed to the point at which traditional blockages and breaks in service (e.g. due to the weather, quarrying and manufacturing problems at source) are no longer regarded as 'one of those things'. Main contractors and clients expect the building materials and supplies sectors to work around this.
- **Building and civil engineering:** the speed at which projects can be completed is the function of initial and continuing site access and levels of resource commitment on the one hand, together with work scheduling, critical path planning and managerial co-ordination on the other. From a marketing point of view, it is a question of speed of returns, client satisfaction and general presentation. Also from a marketing point of view, the more quickly the work is done the better (even though this may not be an operational priority once the work has been awarded). Again, the consequences to those organisations that do not work in this way is a loss of business to those that do.
- **Jobbing building:** gaining and maintaining the reputation for speed of response and work completion are certain to give distinctive reputation advantage. The two most common complaints after all are failure to start (where customers perceive that the contractor has forgotten them); and failure to finish (where, after a quick start, customers perceive that the contractor has forgotten them).

PERSONAL CONTACT AND ACCESS

Customers and clients require personal contact and access for the following reasons:

- **General information:** akin to browsing around a shopping centre, comparing prices, quality, availability, volumes, indications of deadlines and deliverability; this is generally looked for from a wide variety of sources.
- **Specific information:** following a decision to purchase or commission in which the general information is translated into real proposals and client benefits; and again sought from a variety of sources.
- **Detailed information:** following the decision to use the given organisation and indicating the ways in which the work is to be carried out as the precursor to prequalification, tendering and contract award.
- **Contract formation:** agreement and signing which is based on authority as well as information.
- **Contract liaison:** based on a format established and agreed at the point of award and which exists for the duration.
- **Contract delivery:** based on completion to everyone's satisfaction.

In general, personal contact and access are therefore based on a combination of availability, expertise and authority. In specific cases and for specific contracts this has to be precisely tailored. It reinforces the point made elsewhere (Chapter 2, page 31) of the mix of awareness, familiarity and contact; and the need above all to base this on high-quality, expert and continuing personal and professional contact. As well as the key to promotion, it is also in many cases the place of transaction or at least the point at which the transaction is set up.

As with the marketing of consumer goods, the use of place is a part of the demonstration process; and as with consumer goods, at any given time clients and customers will **not buy on the given occasion**. However, attention to personal contact and access has still to be paid:

(a) as part of the process of demonstrating the full range of choice and the qualities and expertise of the particular organisation;
(b) because everyone else is doing it;
(c) to maintain position, familiarity, general knowledge and awareness;
(d) to expand position, familiarity, general knowledge and awareness.

Personal contact and awareness are underpinned by the professional expertise and attitudes of those with whom the client base comes into contact. They are undermined by lack of professionalism and negative attitudes. These may exist in fact, or they may be perceived to exist by the clients because of a lack of positive or desired response from the contractor's marketing staff.

Personal contact and awareness from the contractor's point of view are concerned with devising, designing and presenting the organisation's distinctive expertise as client benefits and presenting them in the most convenient form and way possible. The mix of this clearly varies between activities and organisations. It also varies according to the nature of responsibilities pinned down in specific contracts. It also clearly varies according to the different parts of the industry.

It is therefore necessary next to distinguish between different forms of distribution. This may be:

- **Direct:** between contractor and client – most common in architecture, building and civil engineering, planning and consultancy services.
- **The use of intermediaries and agents:** especially project management, value management, purchasing and supply, estimating and deliveries.
- **Subcontracting:** of specialist work and subprojects; subcontracting also covers parts of workforces; it also covers some specialist and consultancy services.
- **Franchising:** where products and services are branded and where distribution is placed in the hands of franchisees who buy the right to trade under the particular name in returns for guarantees on product and service quality and delivery.

Note

Delivery gap: the immediate problem that becomes apparent is called the delivery gap. This does exist in direct contractor–client relationships where gaps arise as the result of misperceptions and misunderstandings. In anything other than a direct relationship, the problem is compounded by the fact that intermediaries are present.

Whatever form of distribution is used, the purpose is to produce the product or service in the most cost effective or project effective way at the point of agreement (for contracted work) or distribution (for services and materials). There is therefore likely to be a mix of these in all but the simplest of transactions.

The following criteria have to be taken into account.

Location of client base

The key criterion for choosing the place/means of distribution is to provide the best possible access to the client base. Whether mass marketed or specifically targeted, this means knowing and, if necessary, creating a place and/or the outlets required.

One-off projects and products mean creating the point of distribution especially. There are two elements to this – first, creating the place in which the preliminary work is done once the specific properties of the target group have been established; and second, creating the place in which the specific and detailed work can be done as the contract is finalised.

For work based on reputation, the best approach is to develop a place in the segment in which reputation is the key to success.

At the jobbing/domestic end, this means becoming well known in the community, both personally and professionally; and, through the work, building a reputation for good-quality work and customer satisfaction.

At the professional services end, this means carrying out activities that have the same effect and that may include the sponsoring of competitions; research work; feasibility studies; self-started proposals (e.g. for urban regeneration and regional development); published critiques of proposed projects and initiatives.

From whichever point of view, the object is to make the end result known and perceived as coming from an authoritative and expert body or practitioner and this, in turn, leads to the client helping to create the 'place' by seeking the body or practitioner out.

From the point of view of the building and civil engineering sectors, the onus is on becoming a good corporate citizen and making a full contribution to the local community for the duration of the given project. This involves:

(a) attention to site maintenance, appearance and access; traffic management; access to other facilities; where appropriate, compensation for those whose lives have been – and are being – disrupted as the result of the work in hand; and
(b) attention to the needs of the community; provision of facilities for the community (e.g. meeting rooms, guest speakers for local groups); answering any specific concerns; and addressing any opportunities such as education and work experience provision for schools and colleges.

Mass market activities in construction mainly concern the supply side of building and domestic products and materials. The issue for suppliers is therefore to gain the maximum number of outlets through a combination of advertising and differentiation activities, and supported with a distribution chain based on retail supply principles – flexibility of packaging; flexibility, frequency and reliability of distribution; attention to product quality; and attention to the long-term supplier/distributer/retailer relationship (see Summary Box 6.3).

SUMMARY BOX 6.3 Supplier/Distributer/Retailer Relationships

There are two points of view here.

1. **Effectiveness and success**
   The supplier/distributer/retailer relationships are based on mutual long-term profitability and interest. The relationship is spoilt when this balance is lost. This occurs as follows:

   - Where the desired frequency of supply or distribution cannot be assured.
   - Where the retailer can no longer shift previous volumes of the product and makes requests to vary the amounts to the detriment of the supplier or distributer. This may either be: a reduction causing both supplier and distributer to have to look for alternative outlets; or an increase, putting pressure on the existing capacity of supplier and distributer.
   - Where the supplier can no longer adequately supply the volumes necessary to keep the distributer and retailer contented.
   - Where the distributer cannot match the supplier and retailer demands.

   It is also true that as product design becomes simpler, products hitherto reserved for trade markets become more accessible to retail type consumers (above all, the DIY market in this context). Increases in demand for products at this end of the market may therefore affect the long-term future of the commercial and jobbing sector. This also applies to the capability to package base and primary products (e.g. cement, ready mixed concrete, bricks) and provide simple sets of instructions that again enable individual consumers to use them successfully.
2. **Domestic, refurbishment and upgrade products**
   For example, double glazing, fitted kitchens, fitted bedrooms, fitted bathrooms fall into the grey area between retail and capital, and marketing effort has to reflect this. The supplier has to ensure long-term availability of good-quality components – kitchen, bathroom and bedroom fittings; glass, frames and fixtures for the windows that can be cut and assembled precisely. The pressures on distributers and retailers – ensuring their part in the supply chain – remain the same. Fitting the products to their particular use is normally the responsibility of retailers, either because they have the expertise for this in-house, or because they have a regular supply of subcontractors for the purpose.

From this, views of the supply and value chains can be built up (Figure 6.2) and indicate the points at which a mutually profitable relationship is achieved in these aspects of marketing.

Effective supplier/distributer/retailer relationships are diluted when there is failure to address each aspect. Terminal damage occurs where end-user satisfaction is not achieved. This is detrimental first of all to the retailer and quickly knock on to the others.

	Supply		Distri-bution		Retail		Fitting
Supply chain	Raw/price primary materials Design	Packaging	Distance Frequency	Repackaging Advertising Branding	Present-ation Access	Convenience Design Branding	
Added value		Convenience Flexibility Speed		Speed of delivery Confidence		Confidence Enhanced reputation Brand choice	End-user satis-faction Perceived added value

Figure 6.2 Effective supplier/distributer/retailer relationships.

Mass supply activities such as the supply of large quantities of primary materials and structural essentials such as bricks and steel stock to building and civil engineering sites bring distinctive problems. Above all, they include ensuring regular site access for deliveries in ways that the immediate environment and infrastructure can support and without disruption to the rest of the life of the given community. This means ensuring the adequacy of supplies and the delivery fleet. Where necessary, this also can, in turn, mean looking for alternative sources either to complement the prime supply source or else as back-up should anything go wrong.

Agency and consultancy services have to consider the nature of both initial and continuing client demand. The normal initial point of contact is by telephone, usually followed up in writing, and so this has to be supported by a level and quality of initial contact that is sufficient to move the relationship on to the prospect of a continuing professional and profitable basis, as well as to make speedy executive arrangements where necessary. This is supported by technology and electronic communication systems wherever possible.

Service levels

Whatever the nature and means of distribution, the products, expertise and services offered have to be supported throughout the whole process from initial enquiry to project/activity completion. This particular level of service clearly varies according to the nature of the work, priority of the client and the overall profitability of the work, and the early identification of potential and specific problems.

The starting point for this is the quality and level of relationship that are required on the part of both parties. This includes attention to the following:

- desired frequency of contact; required frequency of contact;
- time from order to professional engagement;
- specific client pressures (specific contractor pressures are also important but effective marketing requires that these are kept out of the contractor–client relationship as far as possible);
- availability on an emergency basis; other availability when required/demanded;
- expertise availability and continuity of expertise supply;
- what to do when the service cannot be made available;
- general factors of convenience and access;
- progress reviews and payment methods;
- contractor–client relationship and management responsibility for project/service progress;
- packaging, generic expertise for individual clients;
- review periods and format;
- co-ordination between client and contractor, and contractors and support functions.

The priority and mix of these will clearly vary between different situations and different contractor–client relationships.

Profit levels

Clearly, greater attention is paid to ensure a high profile place in a long-term profitable sector or one that shows distinctive potential. However, all work has to be seen in terms of market development as well as returns on the current project. The ideal is therefore to establish a basic quality of place, distribution and access so that it is possible to maintain and develop the sector whatever its current level of profitability and whatever the future demands of actual and potential clients.

Relationships with distributers, retailers and agencies

This applies to those activities that are not directly controlled – especially suppliers of building products or materials through retail and wholesale outlets; and project management and other specialist expertise that is purchased through an agency. For the former, the onus on the organisation is to control the level, quality and frequency of supply so that the opportunity in the market is maximised and optimised long term. It is necessary to underpin this with regular liaison with the distributers and outlets to ensure continuing priority of place and also as an early warning arrangement where the products in question are either losing or rapidly increasing their markets.

For work conducted through an agency the biggest problems are misperceptions surrounding the initial brief. This includes everything involved – from the nature and duration of the contract, the actual expertise required, support services, to the fee content and structure (and total).

Client liaison

This includes potential client liaison and takes place at a variety of levels. The rules for each are the same however – the development of a continuous familiarity and relationship punctuated by regularised and more formal reviews, the purpose of which is to demonstrate continued and improved ability to match client demand with contractor expertise and willingness – at a profit. The nature and regularity of contact may be varied; though a balance has to be struck when dealing with potential clients between over-pursuing supposedly unprofitable lines of business (especially where a long-term general relationship has not led to work), and ensuring that the contractor organisation is well placed to take advantage of opportunities that may suddenly arise. The mix of clients and especially potential clients and prospects varies and is repositioned constantly according to whether the organisation is working to full capacity or not, its willingness and propensity to expand, and the prospects afforded by new clients.

In this context, Table 6.1 is a useful indication.

The mix of contractor–client liaison activities at each stage shown in Table 6.1 varies between organisations. Apart from anything else, the mix itself is a useful indication of: the extent to which possibilities and prospects are turned into work; the percentage of real work in relation to contacts; the reasons for success and failure.

Conducted properly and professionally, the table is an essential source of information to be used by the organisation in the assessment of the effectiveness and profitability of its current range of activities, product/service/offering portfolios. It also identifies specific points – above all, the point or points at which work is won or lost.

The client liaison process has to be underpinned by resources, commitment and expertise if it is to be fully effective. Failure to do so always gives out the message to clients that this is an unvalued or under-valued area of work, and this is in turn always detrimental to performance.

Above all, clients expect both expertise and authority, especially at the formal review stages. This therefore applies to all expertises and disciplines within the industry.

Table 6.1 *The mix of contractor–client liaison activities.*

Nature	*Purpose*	*Continuous contact*	*Formal review*	*Outcome*
1. Possibility/general contact Request for information Prospect	Introduction, familiarisation, initial discussion	General promotion Personal and professional introductions. Targeted literature if possible	Discussion of client's needs. Identification of extent of potentially mutual advantage	Failure – keep on mailing and general contact list Initial success – move to next bracket
2. Distinctive opportunity/ possibility/prospect	Extensive discussions with contacts Introduction to 'authority'	Presentation of perceived match of client need with contractor expertise Specific factors, e.g rules of competition; CCT	Discussion of possibilities in light of client's projected workload. Establishment of possibility/probability/ certainty of tender/offer	Failure – internal review; external review; review with client. Establish competitive disadvantage; advantages of those successful Success – move to next bracket.
3. Tender/offer	Details of tender Details of nature, volume, quality, duration of work Indication of longer-term prospects	Precise matching of client needs with contractor expertise Addressing specific problem areas and specific factors, e.g. competitive contract tendering	Clear indication of prospects of work Specific constraints Specific problems	Failure – review with client; identifying key factors. Success – move to next bracket
4. Work in progress	Monitor and review progress. Continue to build personal, professional and operational relationships	Identify problem areas, teething troubles. May include a committee structure	Statement of progress. Changes required. May include committee meetings	Relative failure – identify reasons Relative success – identify reasons, prospects for the future
5. Completion	Assess for overall success and failure	Identify problems, causes of success and failure	Measurements against actual progress, aims and objectives	Project/work assessment. Prospects for the future

CREATING A POINT OF SALE

This applies to all disciplines and organisations. The key to effectiveness is:

- recognising the nature of general contact required by the client base for information and general prospects;
- recognising the nature of specific contact required when the client base has an active interest in commissioning work; and
- providing a combination of responsiveness and expertise that meets both.

This is then translated into a combination of facilities and presentation. The main problems with effective points of sale include the following:

- **The perception gap is not fully recognised**. This either causes complacency – e.g. 'they know where to find us if they need us'; or a retreat into general familiarity, e.g. 'everyone knows where we are'. At the other end of the spectrum is information and contact overload, the equivalent in retail terms of carrying a shop or outlet chain which the levels and nature of business simply cannot sustain and which is therefore not cost effective.
- **The cultural perception is taken on trust**. This is seen from two points of view – the industry: 'this is the way that everyone does it' (and therefore it must be right); and the client – 'this is what clients need/want/expect' and which often starts out as accurate and informed, and then loses its currency as other players develop their position.
- **The bunker mentality**. This occurs when an organisation has a long history of success and achievement. The result is that it starts to believe that 'clients will come to us when they need this work doing'.
- **The end in itself**. This occurs when a highly effective point of sale is created but is then not developed as the sector develops and as competitors catch up.

Scale and scope

The scale and scope of the point of sale must consider:

- the range and nature of the contacts made from browsing, general information, individual information through to work requests;
- the number of contacts made broken down as listed previously;
- the frequency of sales and purchases; and this relates to volumes (e.g. one per week, one per minute) and to the percentage frequencies (e.g. 10% of prospects will make further enquiries, and of these 10% will actually make purchase or commission work);
- purchaser convenience;

- strengths and weaknesses of the points of sale chosen as a pointer to future developments;
- strengths and weaknesses of the points of sale chosen by competitors;
- specific technological opportunities and constraints, including the need to provide specific technology and the expertise to use it for prospective customers who make contact (and this includes both physical and electronic contact);
- the current state of the market – expanding, static or declining; stable, shifting and transient; threat of entry; the creation of alternatives; the ability to open up new markets.

Creating a point of sale is often called 'channel matching' – matching the demands of the client base with the properties, qualities and expertise of the contractor so that the desired approach of each is compatible. The process should match the technological, operational and, above all, perceptual factors involved; and should recognise (a) that this is never perfect; (b) the extent of any problem caused by those imperfections; and (c) what actions are ideal, necessary and desirable to remedy these and minimise their effects (see Figure 6.3).

MARKETING TO VESTED INTERESTS

Vested interests – economic, social, political interest groups, community groups and often protest and action groups – have to be recognised, located and addressed. Vested interests are dealt with under the heading of place and location because the reason that they arise is invariably the result of location: the location of proposed activities; the effects on the location (e.g. construction blight, environmental damage) of the proposed activities; the real and/or perceived effects on the quality of life of those in whose area the work is to take place. The specific concerns therefore vary. The principles for dealing with them do not and the following keys are important to bear in mind.

Empathy

Empathy means recognising the effect on the resident and commercial population of having to live in the long-term with the finished project or facility (and with its construction). This means dealing with effects, positive and negative, on:

- **quality of life:** clear presentation of the costs and benefits overall;
- **land and property prices:** which may and does involve underwriting, compensation, guarantees, buy-outs and buy-backs;

PHYSICAL	TECHNOLOGICAL	PERCEPTUAL
• Location	• Telephone	• Convenience
• Appearance	• Fax	• Positive/negative
• Environment	• Computer	• Expectations
• Suitability	• e-mail	• Confidence
→ Accessibility	• WorldWide Web	→ Accessibility
	→ Accessibility	

Figure 6.3 Point of sale: summary.

- **visual aspects:** all construction and civil engineering projects, positive and negative, change the appearance and usage of the land and environment;
- **specific concerns:** these vary from place to place and normally include – light, noise and air pollution; spoil removal; land and environment restoration; traffic disruption; long-term consequences (e.g. the building of a small exclusive estate may set a precedent that opens the prospects of extensive further development).

Respect

Proposal presentations are carried out in ways that respect the general concerns for the locality. Failure to do so simply aggravates those affected. For contentious projects, this always adds fuel to the fire of controversy. Problems of real and perceived lack of respect arise when:

- issues have clearly not been thought through completely or thoroughly;
- costs and benefits (especially benefits) are not what they are deemed to be and are not presented as such to those affected;
- the opinions of the vested interest are dismissed as of no consequence;
- there is a real or perceived secondary or hidden agenda, e.g. the ability of a powerful contractor to break into an area; a politician's desire for a monument to himself;
- there is a problem with priorities, e.g. the council wants a shopping centre, the population wants a bypass;
- unwillingness to address real or perceived concerns, all of which have to be seen from the point of view that:
  (a) they are important enough to those concerned to have raised them;
  (b) failure to do so in all situations is always a mark of lack of respect;
  (c) if the project has been properly planned and thought out the concerns can always be addressed.

Information

Vested interests always want information and this must always be complete. Where it is not, the vested interest always assumes that it is being treated with disrespect, if not dishonesty. Moreover, people always make up the difference between what they do know and what they wish or need to know – and this is invariably negative and detrimental.

The main reasons for not providing complete information are:

- That it is not available – where this is the case, it should always be acknowledged. It is underpinned with a commitment to give it out as soon as it does become available.
- That it is available but is not given out because nobody on the contractor's side has thought it sufficiently important. There is no problem in this case with making it available.
- That it is available but is considered too commercially, economically, socially or politically sensitive for some reason. This information should always be reconsidered with a view to publishing it. There is very little that does not benefit from publication (rather than cause trouble for lack of publication) in most situations. There are very few genuine commercial secrets and the contractor, especially, should be genuinely aware of what these are when they are deemed to exist, and of the prospect of keeping these genuinely secret in the long-term.

SUMMARY BOX 6.4 **Lack of Complete Information**

High Speed Rail Link to the Channel Tunnel: UK and France

This project was due to be completed as part of the wider infrastructure development of the Channel Tunnel project. It is useful to contrast the experiences of the UK and France.

UK

In the UK, the project was met with universal opposition, because it was late in conception. It was not fully supported, especially politically. Benefits were required to be demonstrated before the link itself was commissioned. Moreover, the original route chosen – still widely deemed to be the best – was cancelled at the drawing board stage because it was thought politically, socially and environmentally sensitive. Operationally this route had met all the requirements of high speed rail travel – a flat terrain, long straight stretches and treading a similar route to well-established rail and road networks. It would also have linked centres of population and commercial activity between London and the Tunnel, and therefore with the rest of the European network.

This gave the lead to those in other areas when alternatives were put forward. They knew that the ideal route had been turned down and the reasons for this decision. The perception was that subsequent choices were fudges and compromises, and could therefore be expected to be turned down also provided sufficient protest was made.

The problem was compounded by a total of twelve different routes and proposals being made over the period 1986–91. The universal impression was therefore that politicians, officials and others with influence did not know what they were doing – and this was translated into the community as a widespread lack of confidence, coupled with the belief that, as the original route was turned down as the result of pressure and influence, others could be also.

Other factors that contributed directly were:

- The unwillingness of potential contractors to discuss their project's costs and benefits in any detail. This was made worse by political interests in particular areas gagging the proposers because of the negative tide that was already running against the idea.
- The unwillingness of the UK Department of Trade and Industry or Department of Transport to give any form of lead or direction in the matter.
- The avoidance of any discussion of the principal aims of the project. Lobbies and vested interests concentrated on peripheral issues that became real major concerns – noise levels, site access, construction and civil engineering methods.

None of these would have been top priorities if a clear direction for the project had been established at the outset.

The overall result was that people living on or near all of the proposed routes suffered property blight and therefore material economic loss. The project was universally detested and discredited. To recover any confidence for the project required major marketing effort and economic and political support.

France

In France, the project was early in conception and fully supported. Concentration was on the benefits and not the costs. Disputes therefore arose, not because of the effects on those on the chosen route, but because of the economic loss to those areas that were not on the chosen route. The political consequence of this was that, as well as investment in this project, the French government had to commit itself to providing investment in other transport, infrastructure and commercial development initiatives – especially the coast motorway between Calais and Rouen and regional development grants for the city of Amiens – as compensation for the loss (and the perceived loss) suffered as the result of not being placed on the chosen route. In this case, the vested interests were lobbying to have the project come to them, not against them – yet they still had to be addressed in the same way.

The high speed link between the French Channel Tunnel terminus at Sangatte near Calais and Lille, Brussels and Paris was completed in late 1993. It was designed to be integrated with the rest of the rail infrastructure of Northern Europe, and especially with the TGV (high speed train) networks in France and the long-distance routes connecting Brussels with northern, central, southern and eastern Europe.

Language

Language must be simple, clear and direct – and truthful. Where any of these are not present, the message received is perceived to be dishonest, disrespectful and not worthy of trust. If there are detailed and technical issues to be addressed, then these have to be explained in non-specialist terms and supported by technical documentation that is available for inspection and fervent discussion when necessary.

The problem of language has to be supported by non-verbal communication elements of access and availability so that any pressure group can gain real answers to its genuine concerns (see Summary Box 6.5).

SUMMARY BOX 6.5 **Language**

Pressure groups and vested interests are always encouraged by the use of direct language such as:

- 'I don't know the answer to your question, but I will get you one by – [and give a date]'
- 'I do know the answer to your question and this is it . . .' [and proceed to give the answer in clear, direct and simple terms]
- 'The project will start on [give a date] and be completed by [give a date]'

It is impossible to speak in these terms without reinforcing the message through clear and supportive body language. On the other hand, pressure groups and vested interests are always disconcerted by the use of such phrases as:

- 'It is the view of the inquiry that . . .'
- 'We will have to get an answer from the Minister/council leader/project liaison officer . . .'
- 'That question is beyond the remit of this inquiry'

Each of these phrases – and their equivalent – represents a combination of lack of respect, lack of clearly delineated responsibility and the well-known bureaucratic obfuscation. The message given out is therefore that what is being stated is, at best, bet-hedging; and, at worst, obfuscation bordering on dishonesty. The normal result is to encourage pressure groups and lobbies to extend and expand their range of activities, for example to gain national media coverage, and highlight their particular plight or concern.

Responsibility

Ultimate responsibility for all work rests with the client and this remains true even where responsibilities for dealing with pressure groups are accorded to the contractor or where the effort is divided. This means that any answers given are deemed to carry the full support and confidence of the client; and where answers are not clear, the client can expect further and supplementary requests. Problems arise where:

- Ultimate responsibility is mainly political and therefore not expert.
- Those who are ultimately responsible mistake noise for substance and start to distance themselves from the effects of worthwhile projects and activities. In many cases, this has led to the dilution and cancellation of activities previously deemed necessary and worthwhile (e.g. cancelled airport developments at Manston and Yardley Chase; dilution of the M25 project).
- Those who are ultimately responsible mistake substance for noise, and therefore press ahead regardless, despite the fact that there are real issues to be faced (see Summary Box 6.6).

SUMMARY BOX 6.6 **Pressing On Regardless**

We return again to the legend of the Channel Tunnel rail link on the English side.

On the morning of the Planning Meeting in late 1989, one of the officials concerned realised that he had not carried out his particular activity which was to propose a route from Ashford, through the Kent countryside, to north London. He therefore raced into work early and got himself the largest scale map that he could find. On this he drew a line showing the proposed route, drew a deep breath in a sigh of relief, and went along to the meeting.

His proposed route was accepted as a real possibility, duly published in the local and national press and accorded coverage on the local television station. Only at that point did it come to light that the map that he had used was an old one and that the route he had proposed cut straight through the middle of exclusive executive housing developments on the northern edge of the city of Maidstone which had been built subsequent to the publication of the particular map! The problem was compounded by the fact that rather than apologising for the error and admitting it, officials did their best to justify the shambles.

Effective marketing and presentation to lobbies and vested interests therefore address their concerns from their point of view. The ability to do this is based on recognising the obligation to do it and the level of understanding required. Any project that is supported in this way is much more likely to come across as being worthwhile and of long-term value and benefit than one which is not. Acceptance of this obligation and managing it effectively also gives off a clear impression of confidence and expertise on the part of both contractor and client.

INTERNATIONAL FACTORS

As with all industrial and commercial sectors, the construction industry has become internationalised and globalised over the past twenty years and this is set to continue. The immediate consequence of this is that there is competition for work from companies from all parts of the world for hitherto assured, traditional and domestic markets and sectors. As companies face global competition on their own patches, they therefore in turn have to seek work elsewhere.

It is useful to classify international markets as follows:

- **Steady/static/mature:** Western Europe; urbanised parts of the USA, Canada, Australia, New Zealand.
- **Active:** refurbishment and redevelopment in Western Europe and urbanised parts of the USA, Canada, Australia, New Zealand; infrastructure development in the USA, Canada, Australia; facilities development in Western Europe; urban and infrastructure development in the Middle East, oil states, Hong Kong, Japan.
- **Emergent/pioneering:** ex-USSR, Poland, Czech Republic, Slovakia, Hungary, Mexico, non-urban USA.
- **Emergent/pioneering/active:** China, Vietnam, northern and southern Africa, South America, Mexico.
- **Burgeoning/developing rapidly:** Malaysia, Philippines, Indonesia, Thailand, Korea.

This classification is essential in order to set in context the perceived/received wisdom that the industry's honey pot lies in the area described above as emergent/pioneering/active. In these countries and regions, urbanisation, industrialisation, commercialisation and the upgrading in all aspects of infrastructure are all going to mean vast amounts of work for the industry at large. However, this transition is going to take place over the long-term and this becomes the first implication of international marketing.

International marketing as an investment

A long-term commitment is based on establishing a long-term reputation from which there is a genuine prospect of long-term work. The groundwork has to be carried out first and this may take many years, as witness the number of UK multinational companies that have had a presence in different parts of the world for long periods of time and yet whose main portfolio of work still remains domestic.

Beyond this, there is a strong general perception in all organisations that, because they themselves know their own strengths and capabilities, and because they are well known in their own areas, everyone else must be aware of them also. This is downright dangerous. Whatever their size, companies going into new areas for the first time should assume that the new potential client base knows nothing about them. Acceptance of this view means that it is much more likely that marketing activities will be properly structured and targeted. The opposite of this (see Summary Box 6.7) is that the company assumes that the new market is familiar with them and therefore neglects the groundwork. This is especially true where the particular companies already have carried out some work in the new country and they therefore assume that they are well known. They neglect the fundamentals and the normal consequence is that, while it is possible to continue to pick up odd projects, it is not possible to generate a commercially viable volume of work. The international activity therefore remains peripheral.

SUMMARY BOX 6.7 Common Misperceptions of International Markets

- 'They are waiting to welcome us with open arms'.
- 'They know exactly who we are and what we can do'.
- 'We are miles better at our job than their domestic industry'.
- 'Our construction/architecture/planning/consultancy is the best in the world'.
- 'We can get in there quick, make quick profits and get out'.
- 'We can work substandard for high prices; it is a licence to print money'.
- 'We will get a clear run at the work'.
- 'There's plenty to go round, so we are bound to get some of it'.
- 'We have influence at home, therefore we will have influence over there'.
- 'They will do things our way'.

Behind each lies complacency born of a combination of the certainty of excellence and past achievement with the perception that this will be held in high regard in the proposed market. Worst still, these attitudes indicate a bunker mentality that may be summarised as 'all we have to do to succeed is to go overseas'.

Worst of all is the fact that in many cases the host country and its client base would indeed be only too pleased to use the particular organisation, if only it would conduct its marketing and presentation in ways suitable and acceptable in the particular situation.

SUMMARY BOX 6.8 **Playing Away: Successful International Marketing**

The lessons from Japan

Japanese companies from all industries – and including those that serve consumer markets – have been so successful in gaining extensive business overseas because of the amount of long-term investment and effort that they put into gaining a foothold. They studied the markets; they studied the culture; they studied the patterns of behaviour of those people to whom they were going to be selling. And they presented their products as benefits to the domestic market rather than on their own technical merits or as the result of an obviously unknown domestic reputation (however good that may have been).

In summary, they build their own expertise into the local culture so that what is presented is a combination of: a real understanding of their international client base; a real understanding of the needs, wants and demands of their customers, consumers and clients; an absolute attention to product quality; and delivered by top quality, highly motivated staff.

Any company from the construction industry or one of its disciplines (or for that matter, any company at all) should check its international marketing effort against these features. Lack of attention to any of these always dilutes the effectiveness of international marketing efforts.

Going international

Companies go international for any or all of the following reasons:

- they are being driven out of their domestic or traditional markets by incomers;
- the rules or expectations of their traditional or domestic markets have changed and the company can no longer meet the new set of demands;
- they have excess capacity that cannot be used in the traditional and domestic sectors;
- the domestic market is set, saturated, stagnant, static or dying;
- the domestic market no longer matches the company's expertise;
- the domestic market has a known finite commercial life;
- the overseas market is of the right size, scope, scale and duration to afford opportunities that match the organisation's expertise;
- better margins, quicker returns, long-term stability can better be achieved overseas.

Companies may also find themselves going international as follows:

- **By chance:** for example, the overseas market looks a good bet for some reason. Chance entry may also be achieved as the result of being invited to do the work by an overseas contact. Prospects may become apparent as the result of attendance and participation at professional association and employers' association seminars, or by responding to circulars and publicised invites to tender.

- **Perceived familiarity:** in casting around for work elsewhere it is common to seek some work where there is at least a general feeling of familiarity, e.g. UK companies seek work in Western Europe/the EU, North America, Australia, New Zealand, South Africa, India and Pakistan; US companies seek work in Canada and Mexico. This may be reinforced by the fact that:
  - 'We once did some work there' – whenever that may have been.
  - 'Everyone else is doing it' – when it becomes known that companies are looking overseas, there is a psychological pressure on the rest to do the same, otherwise they perceive that they will somehow be left behind.

Any of these approaches can be – and often are – successful. Each of the above represents a well-used point of entry. The danger is that the few examples where these approaches do lead to genuine long-term prospects are held up as shining examples to the rest of the industry. The greatest prospect of long-term success is achieved by taking a systematic approach.

Systematic approach

The prerequisite to successful international marketing is groundwork. There are four stages:

- **Targeting:** establishing which countries and regions have sufficient potential in terms of market size, scope, scale and longevity to make them worthwhile investigating further.
- **Groundwork:** this consists of extensive targeted market research, the outcome of which is to be the capability of matching opportunity with resources. This is a key commitment and needs long-term resourcing and support. The key result is also the capability to identify, target and prioritise clients and commissioners of work.
- **Familiarisation:** this consists of making contact with the prospective client base; and from this, getting to meet those who have the power, influence and authority to commission work. The initial contact and the person with authority may be one and the same – invariably they are not. Initial contact is therefore likely to be positive and receptive – and influential. By getting to the point of authority, actual requirements can be established and the presentation requirements established also.

The result of this is a real understanding of the environment in which operations are to take place, together with knowing the requirements of the given client base. It is also clear how much work and its value are likely (perhaps certain) to be generated.

The fourth stage runs alongside each of the other three and is a by-product of them. It is the creation and development of a presence and image, ensuring that the new environment gains general knowledge and understanding that a particular company is a serious investor, aiming to establish a long-term productive and profitable presence – and bringing its own expertise for the long-term benefit of the particular country. This means a continuous marketing obligation to the new area, exactly as it is when operating in the domestic market.

Carrying out the groundwork in this way also identifies real problems:

- Cultural barriers: a past history of dealing with companies from the same country as the incomer is invaluable – especially where problems and negative aspects have arisen.
- Specific performance standards.
- Specific political, economic, social, cultural, technological, legal and environmental barriers.
- Language – whenever work is being contemplated overseas, it is always best to engage the services of a bilingual commercial interpreter rather than trusting to employees' command of the language (unless this is exceptional).
- Problems of physical and psychological distance – the ability to control and manage contracts effectively from, in many cases, the other side of the world.
- The competitive advantage of established players – again, this has to be met and beaten if a foothold is to be gained; it is also true (as with the domestic market) that clients will continue to use a contractor with whom they are generally satisfied rather than seeking elsewhere – unless given overwhelming reasons to do so.
- Specific barriers to foreign activities; for example, in some countries foreign incomers are welcomed provided that they enter into joint venture or partnership arrangements with a domestic or more local player.

Once these problems have been identified, then attention can be paid to the ground rules under which successful activities are undertaken in the particular country. Moreover, precise market segments have then to be established; and again, this takes time and investment. The rules of segmentation apply wherever in the world activities are contemplated – and this may require asking international marketers to break an unfamiliar country down by state/region/locality/urban factors/regional factors; to break the population down by age, gender, locality and distribution; and to break the areas of activity down into public projects and works, domestic housing, infrastructure, civil engineering, and commercial/industrial projects.

A point of entry or market penetration has then to be decided upon. The product or service quality has to be absolute. Beyond this, market penetration activities are essential – and this must refer either to penetration pricing (in which a price is set at a sufficiently attractive level to cause work to be initially attracted); product penetration – in which the volume, quality and availability of the product more than meets the capability of other players in the region or sector; penetration quality – in which what is delivered is to a much higher standard of quality than current providers. Beyond this, it is essential to study the marketing activities of the new area. For example, some images and forms of presentation either give the wrong message, or else are culturally (e.g. for religious reasons) unacceptable. Those studying international markets have therefore to make themselves familiar with the ways in which things are done there and to devise means of translating their companies' products and services into these forms of presentation. It is also essential to study the use of language – forms of language acceptable in one part of the world are unacceptable in others, and so on. It may also be necessary to test the acceptability of the direct sales approach and other forms of construction and construction-related marketing.

To summarise, therefore, effective international marketing is about identifying the reasons for going elsewhere; identifying the points of entry; identifying the means of entry; and committing long-term resources and commitment to gaining a foothold. It is akin to starting a new company from scratch – indeed, this is what is happening in terms of the new market.

OTHER FACTORS

Other factors that have to be considered in relation to place, access and distribution are:

- **The nature of the risk:** as established earlier, mainstream design, construction, fitting out and refurbishment activities are high-risk, high-consequence activities. The problem therefore lies in deciding which contractor to use for the first time, whatever their lineage, history, expertise or reputation; in continuing to use them when key personnel have changed (e.g. managers have moved on, specific individuals are tied up in other projects and cannot be released); and in pursuing the same contractors in new environments and locations whatever their history of achievement elsewhere.
- **Core and peripheral business mix:** again, as established earlier, peripheral activities have a useful decision-making function in gaining a

reputation as well as commercial value in new areas. The problem for the client again lies in being prepared to award the contract on the basis of perceived capability, together with the fact that even if the work is 'peripheral' in terms of the product mix, it is still going to be awarded priority and importance by the contracting organisation.

- **Flexibility versus expertise:** client organisations clearly seek a combination of the two and the right mix (whatever that may be) is the ideal. The problem lies in assessing the balance of capability with willingness according to the demands of the given situation. The balance tends towards expertise in areas of high technology, precision and total quality; and tilts towards willingness and flexibility where these aspects are less absolute.
- **Service levels:** it is increasingly the practice for these to be guaranteed, or at least a set of boundaries and rules to be established. The most common forms are as follows:

    - service level agreements – identifying the main features of service required and expected so that a clear indication is established;
    - service level contracts – identifying and establishing all the features of service required and expected, and formalising them as part of the total contract;
    - competitive contract tendering – in which anyone wishing to tender for work is expected to enshrine certain key features as a prequalification condition. The most common features of CCT are union recognition agreements, training agreements, rights of public and vested interest access, environment management and protection. However, anything may be included by the client.

- **Vested interest management:** in which the client requests that the contractor will play at least a joint role in the management of influential and/or legitimate public concerns and other interest groups. This may be a general partnership agreement or specific areas may be agreed and assigned to be dealt with by each party.
- **'Any other duties':** contractual arrangements increasingly include the riders – 'to act at all times in the client's best interests' and 'to identify and resolve such problems that may become apparent during the duration of the contract'. While there are clear financial and ethical considerations here, it indicates the emphasis of the contractor–client relationship.
- **Relationship management:** much of the place, access and distribution aspects of construction marketing are concerned with relationship management. Primarily this is the management of the direct relationship between contractor and client. It also includes generating positive and confident relationships between:

(a) the contractor and the client's governors, resource controllers, political interests, public interest, vested interests and other pressure groups; and where appropriate, the end-users of the complete facility;
(b) the client and the contractor's key supporters – including specific, technological and expertise provided by agencies and consultancies; specific subcontractors; and contract backers;
(c) the management of the relationship between contractor, client and media – the nature and interest of the media vary according to the nature of work; there is invariably some interest, however, whether specialist, technical, general, local or national;
(d) managing the relationship between the contractor, client and community with a special reference to generating a good level of respect and value for the matter in hand.

- **Environment management:** this varies between disciplines. Whatever the case, it remains a key and current concern of the marketing aspects of construction management.
- **Building and civil engineering:** this concerns environmental maintenance and increasingly, refurbishment and restoration at the end of the contract. It also includes health and safety, appearance and other perceived quality aspects of site and work in progress management. It may also include attention to the relationship between the site and its environment so that disruption to the other activities of the locality are kept to a minimum and any specific work factors (e.g. mud, dust and grime) are recognised and cleaned up.
- **Architecture, design, planning:** this concerns the understanding of the impact on the environment of particular proposals, projects, initiatives and designs. It involves having a distinctive view of the costs and benefits of each.
- **Building materials:** the two concerns here are maintaining a positive and agreeable environment at the point of sale or distribution, including the management of mud and dust; and also the effects on the environment caused by deliveries in and out.
- **Consultancy and professional services:** the additional concern here is to be able to learn the given environment and its specific pressures, often in a very short time. There is the additional problem of seeing the application of the expertise in its environment and understanding the wider impact of particular proposals and choices.

  For example, a drainage consultancy may find themselves recommending two alternatives – one of which is a high-cost proposal including full water management, disposal and channelled access watercourses with full provision for storms and floods; and the other an adequate, cheaper alternative that will do an efficient job, all things being equal. The consultancy then has to live with the consequences of the choice that it has made – the higher cost alternative may damage the relationship with the main contractor; while the lower cost choice may cause questions to be directed at the consultancy if there are long-term problems with future capacity.

- **Corporate citizenship:** this takes a wider view of environment management. The starting point is that of 'the organisation in its environment'. The organisation takes an empathetic and enlightened view of its position, responsibilities and obligations over the medium to long term. This is partly because it is good general management to do so and partly because it is good business (and therefore marketing). Clients are much more likely to return to a contractor for future work if the organisation concerned has taken a fully enlightened view of its responsibilities.

Above all, the effect of building and civil engineering work on the environment is a prime, current and continuing concern in all parts of Western Europe, North America, Australia and New Zealand; and it is certain to become more of a concern in the other parts of the world in the near future (see Summary Box 6.9). The more knowledgeable and enlightened the view taken, therefore, the greater the understanding of both specific and general environmental pressures.

SUMMARY BOX 6.9 The Construction Industry and the Environment in the Third World

The rush for civilisation

One of the main features in marketing and contractor–client relationships is understanding the conflicting pressures of the rush for civilisation and the environment.

It is very easy to take a supposedly enlightened Western view of this. This is emphasised by the presence of global forums such as the United Nations Conference on Environmental Protection which, in March 1997, set specific recommendations for limitations on deforestation and greenhouse gas emissions for every member nation.

On the other hand, there are overwhelming pressures to urbanise and produce a version of Western civilisation, with all the benefits that this is perceived to bring. Such pressures are founded on the development in the Far East and ex-Communist Bloc of industrial nations, producing goods for Western and Japanese multinational companies. This process has transformed these economies from agrarian to industrial in the Far East, and from command to demand in the ex-Eastern Bloc. In turn, this has resulted in the need for housing, infrastructure, sanitation and communications networks. There are also political pressures for the work to be carried out quickly.

The result is often that the work is carried out without due concern for environmental protection and, in many cases, this is not even recognised as important. Where Western interests and influences have expressed concerns, the Eastern and Far Eastern clients have treated these views with suspicion, often seeing themselves as being reined in, rather than giving themselves the chance to develop fully.

There is, therefore, a moral choice when marketing in the Third World, and it may (and does) extend in some cases to standards and quality of work, as well as to environmental impact and aesthetics. Those seeking work in these places have therefore at least to be aware that this can occur. They may also be asked – even pressurised – to use materials with which they are not fully familiar, and in which they do not have full confidence.

They may be asked to work to designs that do not accord with their accepted ways of doing things, or meet familiar usage and safety standards. In March 1997, there was a 'Towering Inferno' type fire in the centre of Bangkok, and 20 people lost their lives. In spite of warnings by the contractors at the time of construction – that the building was both inaccessible and too tall for the city fire service if there was a fire – the project was agreed. There is currently a legal wrangle over responsibility, and this is likely to lead to prosecutions by the city authorities and the Thai government.

7 Promotion

INTRODUCTION

The construction industry is concerned with the following forms of promotion:

- the promotion of its capabilities and expertise to clients and potential clients;
- the promotion of its finished products to the community and society at large;
- the promotion of professions within the industry to each other; promoting specific general and continuing interrelationships and confidence between architects, contractors, quantity surveyors, planners, other consultants, civil engineers and subcontractors;
- awareness raising and image building; general presentation of information in the most advantageous ways;
- brand building and reinforcement – the creation of the equivalent of 'brand' around the name of the company in question;
- the promotion of general confidence, public sympathy and support for activities.

All organisations are also generally concerned with presenting themselves as durable and dependable to both customers and staff, and to other stakeholders as one of the foundations of general confidence.

These are additional to the general concern of all organisations in every sphere with presenting and promoting themselves in the best possible light, accentuating their strengths and distinctive capability in ways expected and recognised by their markets and customers, and potential markets and customers.

The overall purpose of promotion is:

- building, developing and enhancing reputation and confidence;
- presentation of achievements, capabilities and expertise;

and relating the two to the needs and wants of customers and potential customers. The key is to use the language and media – especially the language of the receiver (both clients and potential clients, and also within the industry); and to develop media within organisations so that expertise and capability are shown in the best possible light in the eyes of receivers.

In order to promote the industry and its disciplines and activities effectively, it is therefore first necessary to recognise customers' and clients' expectations. The main aspects are:

Table 7.1 *Differing patterns of consumer goods and construction consumption.*

Consumer goods	*Construction*
Short lead times	Long lead times
Instant purchase	Considered purchase
No/few consequences	Enduring consequences
Self-driven/individual driven	Commissioned by groups
Cash/credit payment	Capital payment
Easy replacement	Difficult replacement
Low opportunity cost	High opportunity costs
Low durability	High durability
Instant satisfaction	Enduring satisfaction
Instant benefits	Enduring benefits

- the ability to meet deadlines;
- the ability to deliver the agreed design/product/project to the agreed cost;
- quality assurance, durability;
- after-sales service and support.

These elements are then translated into promotional activities suitable to the construction industry.

It is next essential to recognise the fundamental differences between consumption patterns of consumer goods and construction 'consumption' (see Table 7.1).

The following elements also impinge on the promotion of construction:

- **Political factors**: have to be taken into account in most circumstances. For example, all construction projects affect the environment. The effect of this has to be considered in all circumstances therefore, and part of the promotional activity is concerned with addressing this. The matter is most critical where there are both long lead times and long project times for major initiatives – such as the Channel Tunnel itself or the rail link. It also happens to a lesser extent with housing developments (for example, the transformation of areas through the construction of high rise flats or other intensive residential accommodation; and to a lesser extent, the development of 'exclusive' residential accommodation in rural and/or green belt areas). It also occurs in many cases with the refurbishment of existing buildings. Whatever the case, promotion must address the particular political drives and resistances so that the benefits of whatever is to be undertaken both outweigh the drawbacks, and are presented as such in ways acceptable to the population.
- **Public acceptance and colloquy:** are in any case, absolutely critical to the acceptance of construction projects. Projects may be designed with the purpose of producing all sorts of benefits to the community

in which they are to be sited; this is no use if the public affected do not perceive these benefits or their value. By the same token, it is perfectly possible to produce projects that bring little or no benefit to communities; and, yet, the promotion activity has been carried out in such a way that these are accepted anyway! The standpoint taken is therefore that not only must there be benefits accruing from construction activities, but that these must also be presented in ways acceptable and sympathetic to those affected.

Consumer goods promotion concentrates on the capability for instant satisfaction, convenience and the ability to pay. This is based on analysing the propensity to spend, spending and consumption patterns of individual consumers and consumer sectors, and segments targeted. Benefits are presented as **tangible** – use and value of the particular item and the benefits that accrue from using or owning it; and **intangible** – association with the lifestyle, images and other brand properties.

The results of the effective promotion of construction activities are the same. The tangible benefits are the use and value of the particular facility, drawing, proposal or project, and the benefits to be accrued from having it commissioned and built. The intangible benefits are associated with the finished facility, amenity or building, and its enduring benefit to the community. Effective brand building is carried out through a combination of medium and long-term reputation enhancement, based on consistent production of good-quality buildings and facilities. Promotion activities exist also to reinforce organisational credibility and to build and reinforce customer and client expectations that are to be met as the result of dealing with the given organisation.

This is further reinforced by the fact that few construction products are identical. This is a fundamental distinction from the production of consumer goods, where production runs (e.g. of aspirin, televisions, canned foods) often run into millions of units, all of which are identical. Promotion emphasis has therefore to be targeted at the virtues of the uniqueness together with the capability to produce it, as well as general confidence (see Summary Box 7.1).

REPUTATION AND CONFIDENCE

The reputation of companies in the construction industry is built as follows.

Industry specific factors

- The reality of their finished products and services which are there for the public to see and use; their useful life; the nature and amount of satisfaction that they bring.

SUMMARY BOX 7.1 **High Speed Rail Link**

The public reception to the high speed rail link between the Channel Tunnel and the networks of the UK and France contrast utterly. In the UK, there was great resistance to the entire project; in France, the towns and cities of the northern part of the country lobbied furiously to have it close to them.

The reason for this lies in the way it was promoted on each side of the Channel. In France, the project was seen as bringing benefits to all the communities on which it was to touch. It was perceived to open up industrial and commercial possibilities, to improve distribution of products and services, to improve general accessibility, and to bring prosperity to those towns fortunate enough to be linked by it.

In the UK, none of these benefits was spelled out in ways acceptable to the general public – and, especially, those parts of the population that would be directly affected by it. Attention instead focused on construction and environment blight. This was exacerbated by the inability and/or unwillingness of those responsible for deciding the route to come to a firm decision. Indeed, at one point, there were no fewer than 12 proposals for the route lodged with the various authorities. This was, in turn, compounded by what came to be known as 'link blight' – which above all resulted in the inability of those situated alongside or near to any of the proposed routes to sell their houses; and houses lost value as the result.

In the UK, this was compounded by the adverse media coverage that resulted from the uncertainty. Every slight uncertainty was seized upon. Every item compounded the uncertainty. Whether the project would bring benefits to the community became a side issue; the point was that public confidence in the initiative was lost and the overwhelming drive therefore became to ensure that the project was not undertaken.

- The extent to which the finished building blends in with the environment; enhances the environment; and contributes to the environment.
- The additional value and satisfaction brought to people's lives by buildings and amenities.
- The quality of life that they underpin; other related matters of public status and perception (e.g. by living in place *x*, by working in place *y*).
- Wider public perceptions influenced again by utility, public reception, press and media reaction.
- Appearance of the building, facility or amenity and the extent to which it complements and blends in with the rest of the built environment; the extent to which it is timeless or faddish.
- Effective functioning by serving its intended purpose well.
- The variety of uses to which it can be put and extent to which it can be changed to accommodate different uses.

General elements

- The variety and volume of work carried out.
- The nature of contractor–client relationships.
- The nature of inter-professional relationships between construction disciplines, clients, potential clients and the public at large.
- Specific attention to cost, added value, quality, functioning, aesthetics and deadlines.

Organisational factors

- Every communication that takes place.
- Head Office reception – appearance and demeanour.
- Site reception – appearance and demeanour.
- Site security.
- Tendering and pretendering presentation.
- General conduct and demeanour.
- Conduct and demeanour in public (e.g. at public inquiries).
- Conduct and demeanour when handling adverse stories and publicity.
- Handling of clients on a daily basis and the forms of liaison that are established.
- Style of response to customers, clients and enquiries.
- Style of approach to customers, clients and enquiries.
- Meeting deadlines.
- General presentational aspects.
- Answering the phone, taking and passing on messages.
- Speed of response to queries and enquiries.
- Road manners of drivers; manners of contractors; manners of subcontractors.
- General ambience of sites.
- Corporate attitudes.
- Managerial attitudes and behaviour; staff attitudes and behaviour.
- Acceptance of corporate responsibility.
- General media coverage of the companies.

Organisation information

- Annual reports.
- Brochures.
- Advertising (if any).
- Use of media, press relations, story placement.
- Public relations activities.
- The use of professional and technical journals.

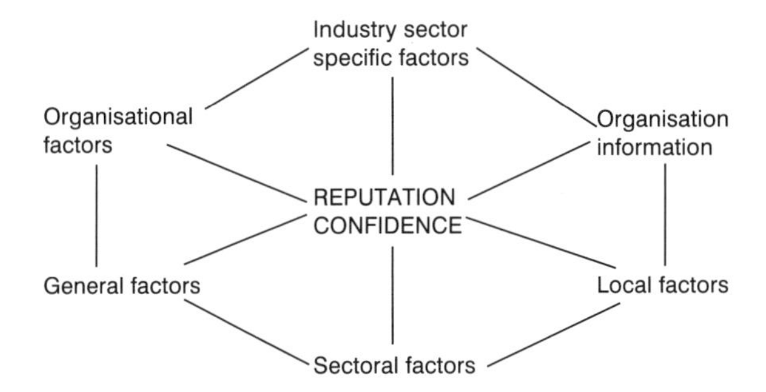

Figure 7.1 Building and promoting reputation and confidence.

Local and sectoral factors

- Small and local operators in all construction disciplines rely heavily on local reputation – word of mouth, appearance of finished product, operator reliability, quality and perceived quality of work, and working to budget.
- Organisations and practices that concentrate their efforts on limited fields of activity must promote a distinctive capability and advantage for themselves and to the sector. As stated elsewhere, this is likely to be a cost, quality, deadline or expertise advantage. The essential factor is the presentation of this in ways with which the client base is sympathetic. For example, one building or civil engineering company specialising in drainage may guarantee its work for a period of years; another may guarantee to have the work carried out to satisfaction within shorter timescales than any other company – this gives potential clients a clear choice. A landscaping practice may establish itself on the basis that it undertakes to have removed any toxic or lesser waste discovered during the course of site preparation.

The foundation of all successful promotion therefore is the ability to assimilate and combine these elements, and present expertise and achievements in ways expected and demanded by client bases (see Figure 7.1).

PROMOTION OBJECTIVES

The purpose of promotion is to:

- support the activities of the company, and present them to the best possible advantage in the eyes of clients and potential clients;
- generate and reinforce more nebulous concepts of reliability and confidence;

- present the company's distinctive features and expertise in the language of clients and potential clients, as well as to its own best advantage;
- assist in the process of meeting customer and client expectations.

The media available to the construction industry, in common with all others for the purposes of promotion, are as follows:

TV	Customer liaison
Radio	Sponsorship
National press	Word of mouth
Local press	General company presentation
Trade press	Public relations
Billboards and hoardings	Promotion through direct mail
Brochures and sales literature	Products and name placement
Personal sales	Other publicity

The following are also available specifically to the construction industry:

- the development of a brand image around the company name, logo and the facility's appearance;
- the development of positive images as the result of current, recent and historic activities;
- company boards at project gates and entrances;
- photograph, television and video presentation of work in progress;
- use of trade fairs, exhibitions and shows;
- editorial/documentary on specific problems and issues;
- academic and industrial published research;
- lobbying and networking;
- competitions;
- the activities of professional and industrial associations (e.g. CIOB, ICE, AA, RIBA, RICS, IStructE, RTPI);
- other activities such as the establishment of an overseas branch, or the acquisition of a company in a different sphere of the industry (e.g. the acquisition of a house building company by a civil engineering company) also contributes to the general promotion profile.

Key factors in the promotion of construction

Promotion activities in the construction industry must take account of the following:

- Presentation in ways that meet the needs and expectations of all clients and their stakeholders.
- Purchases and commissions are infrequent and unique.

SUMMARY BOX 7.2 Edge of Town Facilities and Green Field Sites

There has been much political debate and general concern over the continued expansion of out-of-town and edge-of-town shopping and distribution facilities. It is perceived that these are detrimental to the continued prosperity of city and town centres and that their continued expansion and increased usage will result in blight and decay in the centres.

The reason why they are so successful is because their benefits have been presented so effectively. They are easily accessible. They provide a wide range of goods and services. Construction promotion has been linked to consumer promotion and benefits – those benefits to be gained by convenience and access together with the ready availability of those things that consumers want to buy. This has enabled the owners and operators of these facilities to go on to generate both increased consumer bases and enhanced satisfaction through the provision of (for example) cafeterias, petrol stations and children's play areas.

- The price of the contract and what is covered by that price.
- The deadline required.
- Quality assurance aspects of the finished product and any reference to after-sales service.
- Different forms of presentation and promotion are required for different markets – for example, in design and build, industrial estates and parks contracts have to be addressed in vastly differing ways, though the fundamental principles – of gaining and keeping confidence – remain the same.
- The nature of particular contractor–client relationships and the ability to recognise the key features. For example, time, quality and cost drives in the design and build for out-of-town superstores; the reconciliation of public and political interests with demands for quality and value in the pursuit of public sector projects – have each to be recognised.
- Supporting and developing existing relationships between construction professions, contractors and clients, and the industry and the public at large.
- The vagaries of the wider buyer–seller relationship. Again, this is especially true in public and overseas projects that are subject to political whim and instant changes to specifications. Where such vagaries are important or likely, and where companies have many dealings with public bodies or are heavily dependent on commissions from the public sector for an effective and profitable body of work, and where this is a key contributor to the future of the organisation, these considerations must be taken into account in the forms of promotion and presentation adapted.

Specific problems

Promotional activities have to address particular issues specific to the construction industry.

Public opposition

This can be both initial and continuing, especially where major projects are concerned, and is at its most extreme where what is proposed is to cause:

- serious disruption to people's lives (e.g. closing off town centres to traffic to create pedestrian precincts or centre of town shopping and leisure facilities);
- serious disruption to the environment where no benefit is seen to accrue as the result, as with some edge-of-town and rural developments and road schemes (especially motorways and bypasses – see Summary Box 7.3).

SUMMARY BOX 7.3 Contentious Projects and Promotion

Eurotunnel

When Eurotunnel won the contract to build the Channel Tunnel, the company went to a great deal of trouble to gain local public support and sympathy from those who were to be most affected by the development. The company arranged speakers for anyone who wanted them – and this included schools, colleges, professional associations, local companies, public services and also groups such as the Women's Institute, Townswomen's Guild, the local Chambers of Commerce and church groups. Helplines were established and were always made available. The company produced a range of free pictorial and easy-to-read and understand literature for distribution upon request. It opened an exhibition centre with library and archives, which were at first free, and this included models of the whole facility, the history of the project, the development of technology and mock-ups of the underground activities. This was (and remains) accessible to both individuals and parties, and now includes full conference facilities. The company also always undertook to face the press and media on any matter at all; and undertook also never to avoid responsibilities or acceptance of blame when mistakes were made.

The initial investment in promotion was of the order of £2 million. The company's total investment in this part of the promotion over the period 1986–96 was approximately £22 million. At a total project cost of £11 billion this represents expenditure of 0.05%.

Newbury

The A34 bypass at Newbury, Berkshire, was commissioned by the Berkshire County Council and the Department of the Environment in 1992 with the purpose of taking traffic away from the town centre on what was (and remains) a major north–south route. As with the Channel Tunnel, there was local opposition. In this case, the positive benefits were never demonstrated. Neither contractor nor client undertook the necessary promotion and public relations groundwork. The area of outstanding natural beauty affected became (and remains) a battlefield – between those trying to get the bypass built and those who wish to preserve the countryside.

Information – promotion – itself has become a matter of campaign. Those opposed to the development have gained substantial television, radio and newspaper coverage for their activities. They have been able to present themselves as underdogs and therefore in a sympathetic light. Protestors were able to produce hard information supporting their case from a variety of points of view, and one of the most notorious examples of this was a memorandum documenting the fact that the bypass would increase the average time saving for traffic over the five miles (8 km) in question by a matter of 2 minutes only.

As with all similar cases, failure to promote the given points of view leads to a perceived wider feeling of lack of confidence – and fertile ground for any opposition. The stand-off that has ensued is both adding to the cost of the project in hand and is certain to affect the contractor's future standing, and this far outweighs the cost and value of engaging in the proactive promotion. The longer this is allowed to persist, the greater the feeling of anger and alienation on the part of those affected by the project.

The long term

The promotion and especially the public relations effort must be capable of sustaining public confidence and acceptability for the duration of the given project; and this remains the case whether it is a relatively small matter taking weeks only, or a major infrastructure project the duration of which is many years.

Real and perceived responsibility

It is often the case that a contractor receives publicity (especially adverse publicity) for reasons outside its control. This occurs when projects and activities are held up by adverse and continuous spells of bad weather. It occurs also when a contractor's work is affected by the activities of the client (see Box 7.3 again – the Newbury example). It is therefore essential to recognise in advance where potential and actual problems are likely to arise and, as far as possible, to gear up a promotion and public relations facility to counter these or at least to address the concerns of those affected.

SUMMARY BOX 7.4 Conception to Reality: London Docklands Development Corporation (LDDC)

The LDDC was created in 1981 to refurbish and regenerate the Docklands area of the eastern part of London which had fallen into decay and disrepair as the result of the old dockyards ceasing to be used commercially.

Over the period of the 1980s a spectacular new development took place. The design and presentation used all the latest materials and concepts. The presentation to the public was effected in such a way that it was perceived to be superb. The finished project was to provide 150,000 jobs for an area of considerable unemployment; furthermore, this was to be condensed into an area of two square miles (5 km^2).

At the end of the first stage of the project (1986) reactions and responses continued to be favourable. Those moving out to work in the Docklands from the centre of London responded favourably to the new modern environment and the general ambience of the development. However, as more people moved into the development, responses became less favourable and, by 1989, the conception of the project was called seriously into question.

The main problem to be addressed was access. During the course of the first phase of the project, a light railway system (The Docklands Light Railway) had been constructed and this provided adequate access for residents and for the initial workforce. However, it had not the capacity to transport large numbers of workers around, especially during the rush hours; nor was it connected to the rest of the London transport system. Above all, there was no major railway terminus.

Over the period 1990 to date, the problem has at least been partially addressed and the benefits of the development are seen to be more enduring.

It is a useful illustration of the development of benefits along the 'conception to reality' process. The initial benefit was regeneration; subsequent benefits included enhancement of working life, urban and commercial regeneration, and infrastructure development. It indicates the broadest view required at the conception/design/planning stage and some of the ways in which this may be expected to develop over the course of a project. It also indicates the promotional activities necessary in support of such schemes.

Blight

This comes in six forms:

- **Noise:** certain to be an issue where night work is carried out. It may also be an issue at all times, if activities are taking place in residential areas. It refers again to both noise generated by the project and also by the traffic serving it.

- **Light:** certain to be an issue where night work is carried out; and exacerbated where this is close to residential, rural and other non-commercial areas.
- **Dirt and dust:** caused by the activities themselves and, again, exacerbated by specific factors such as wind and rain.
- **Traffic:** especially contractor traffic, lorry and rail flows, and workforce comings and goings.
- **Work in progress:** the visual aspect of operational construction sites is normally negative and this is often compounded by security fences and hoardings. This form of blight is also increased where there is disruption to the lives of the community – for example, where roads have to be coned off for periods of time; where there are features such as scaffolding that extend into the public domain as in the refurbishment/repair/replacement of building frontages.
- **Social**: where a contractor brings in a workforce from a different location or culture; or where labour is hired in the locality at premium rates, causing disruption to the local labour market.

Integration

A critical part of all construction public relations is the capability to demonstrate the benefits that the proposal is to bring; and that the building or facility created is fully integrated with the existing built environment (see Summary Box 7.5). This is much less of a problem where the benefits are overt, short-term or otherwise self-evident. The promotion emphasis is on the total enhanced quality of life and environment produced as the result, as well as on the distinctive use of the facility or project; and the balance is emphasised one way or the other according to the nature of the project (see Summary Box 7.5).

PROFESSIONAL GROUPS

For the professional groups within the industry there are specific issues.

Architects

- **Reputation:** especially where the practice has produced and built controversial or contentious projects and facilities. It also occurs where architects get pigeonholed by potential clients into specific areas; and this works to their advantage so long as work continues in the area, and against them when this work finishes or diminishes. Architectural promotion is therefore concerned with both maintaining the

SUMMARY BOX 7.5 Integration: The City of Milton Keynes

The city of Milton Keynes devised and ran a marketing campaign in the late 1980s and early 1990s in which it promoted itself as a wonderful place to live and work. The images used were extremely powerful and covered all of the positive aspects indicated. Those included were:

- peace and tranquillity;
- safety and security, both of housing and of the roads, and reinforced by pictures of children riding their bicycles;
- gatherings of people for holidays and festivals, and to watch things like balloon races;
- pictures of shopping and work areas giving the impression of modernity, stability and prosperity, and also comfort and security;
- access, including the wideness of the roads and the inter-city rail link;
- a close yet comfortable relationship between living and working;
- presentation of the position of the city at the centre of the UK with easy access to London and everywhere else (it is actually halfway between London and Birmingham);
- the use of sunshine in all the pictures to reinforce each of the positive messages indicated.

The end result was, and remains, an extremely powerful and successful campaign, fully integrated, both in the general promotion of the city, and with the specific purpose of attracting business, commercial and domestic interest.

existing position and reputation, and also developing this with other sectors.

- **Design to completion:** based on the ability of builders to build what has been designed. This also has reference to the sensitivity of designs, availability of materials, cost of materials, quality and durability.
- **Design durability:** the mix of fashion and fad with facility, endurance and quality.
- **Project/facility appearance:** there is a fine line between creativity and controversy. It requires that the architect takes on a broad and (as far as possible) rational view of this, and undertakes specific promotion and communication activities where controversy is likely to arise.
- **Response to adverse stories and reputations:** generated by media critiques (and these are often uninformed or led by those not part of the industry) of recent and current work. However ill-informed, stories must be dealt with in public in the arena where they are being presented.
- **The value of professional competitions to all those at all levels of seniority and expertise of practice:** this develops expertise, creativ-

ity and imagination; and it also promotes and enhances the knowledge and reputation of the organisation among competition setters and in the wider industry.

- **The place of current and recent work as advertiser of the distinctive expertise of the practice:** for this, an understanding of the use of materials (as well as appearance) is essential. There are also potential problems outside the architect's control where builders run into operational difficulties, and these also may have to be faced.

Planners and planning services

- **Technical knowledge:** a key feature in the promotion of any planning enterprise is the range of technical knowledge the practice brings to particular situations. The broader and deeper the range, the greater the initial confidence; and this is reinforced through the ability to demonstrate (e.g. through brochures, trade press features) the range of expertise and work quality.
- **Environment sympathy:** this is highly subjective and at the mercy of the whims of public opinion and again negative presentation.
- **Empathy:** viewing the finished project from the point of view of the area in which it is located is essential. A broad general understanding of the vagaries of public opinion is also required, the speed at which this can change, and the reasons, whims and prejudices that drive it.
- **Final usage:** the ability to project and present the future facility usage in terms of customer and other people flows; possible and potential alternative uses; and any necessary maintenance activities arising as the result.

Quantity surveyors

- **Access to materials:** as and when required; quantity/volume; quality; alternative sources; the ability to handle crises; the ability to handle variations in orders.
- **Scheduling:** the ability to deliver materials on site when required; attention to quality and volume; access to alternative sources; use of schedules; just-in-time factors; crisis management.
- **Deadlines:** ability to meet deadlines; prioritising; site access, e.g. transport; management of variables, e.g. traffic and railway hold-ups.
- **Cost:** contents of price; price/quality/value/volume/time mixes; price content; value for money.

Builders

- **The utility and value of the finished product:** and this may extend to the effective total life of the facility, other potential usage, refurbishment, upgrading and restoration.
- **Integration with the immediate built environment and the wider environment:** and while some of this is certain to be outside the builder's control, the builder nevertheless is going to have to take some responsibility. Any adverse publicity is certain to reflect to some extent on the building company.
- **Disruption caused by noise, dirt, light and increased volumes of traffic:** the rerouting of traffic and pedestrians; disruption caused by traffic bottlenecks that result from the activities.
- **Behavioural aspects of site work including specific standards of behaviour and attitude:** both towards the work itself and also in relation to the immediate public and environment.
- **Promotion of building services:** whether, for example, the job includes completion, fitting out, after-sales service, maintenance and refurbishment; the capability to do this; distinctive capabilities in these and related fields.
- **Presentation of the capabilities of materials:** in the language of clients.
- **Attention to quality/cost/durability in specific cases:** often presented in terms of client value.
- **Contract compliance:** where, for example, contractors are required to recognise trade unions, hold distinctive organisational qualifications (e.g. BS 5750, ISO 9000); undertake and produce distinctive staff training programmes; pay particular attention to health and safety at work; and other aspects related to personnel, staff and industrial relations management. More widely, contract compliance may also include reference to penalty clauses, liaison with client bodies (especially where public sector projects are being undertaken), and relationships with other special interest groups.
- **Lobbies and vested interests:** especially where the project is of a contentious or controversial nature.
- **The management of building projects:** concerned with promoting attitudes, values, behaviour and standards among both staff and stakeholders; with progress; and with effective reporting and liaison relationships with clients. It may also be necessary to adopt the media/PR function for the contractor or total project.

Consultants

- **Range of flexibility of expertise, and the ability to apply it in all possible situations in the field:** this includes the emphasis of dis-

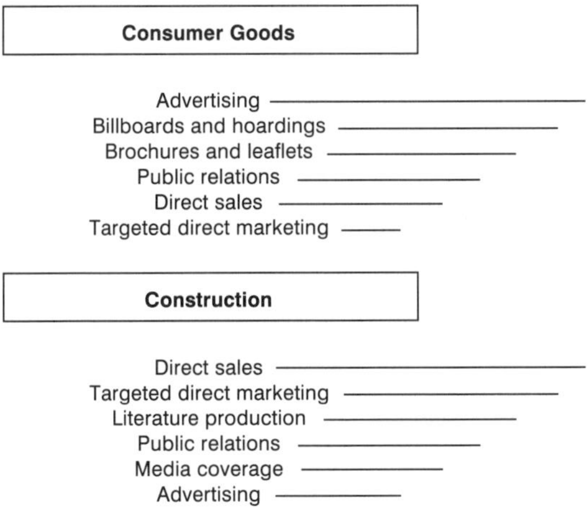

Consumer Goods

Advertising
Billboards and hoardings
Brochures and leaflets
Public relations
Direct sales
Targeted direct marketing

Construction

Direct sales
Targeted direct marketing
Literature production
Public relations
Media coverage
Advertising

Figure 7.2 Promotion activities: construction and consumer goods compared.

tinctive expertise, specific achievements, high profile activities and projects; as well as behavioural capacities of flexibility, dynamism and responsiveness.

- **Relating expertise and capability to distinctive customer demand and requests:** this normally requires the capability to present expertise in a range of different ways.
- **The distinctive contribution to be made if consultants are engaged:** whether it is to resolve problems, add value or enter new fields.
- **The presentation of the capabilities of materials:** in the language of customers and clients.
- **Attention to the quality/cost/durability mix:** in individual cases.
- **Attention to other distinctive client requirements:** especially deadlines.
- **Contract compliance:** in the accepted sense of ensuring absolute standards of human resource management (see Builders above); the general ability to comply with customer requirements and to be flexible enough to accommodate them.
- **Promoting a reputation for high quality:** and a positive approach to work.

PROMOTION ACTIVITIES

Because of the disparate nature of the construction industry and the disciplines that operate within it, and because of the uniqueness rather than the multiple identity of the products generated, and taking into account the vagaries of the industry as a whole and the professions that operate within it, it is necessary to distinguish between construction promotion and that for consumer goods, (see Figure 7.2). Note that effective consumer goods promotion depends on quality and volume of advertising. Effective construction promotion depends on high-quality direct sales and targeted direct marketing activities.

Promotion of the construction industry may be subdivided into primary, secondary and tertiary/supportive activities.

Primary promotion

Primary promotion takes the following forms.

Direct sales

This is targeted at clients, potential clients and others capable of using the distinctive expertise on offer. It involves opening up client face-to-face contact with those responsible for commissioning work and inviting tenders and presenting the distinctive expertise on offer in the client's best interests.

An essential part of the direct sales approach is:

- identifying the right companies and the right people within those companies;
- recognising the nature of sales calling required by customers – what is achieved by the sales call; the extent to which it is genuine selling (as distinct from customer liaison or other forms of general promotion);
- identification of strengths and weaknesses existing within the company's full range of marketing activities – especially product, capability and expertise ignorance;
- identifying those presentational factors that enhance a company's reputation and those that diminish it.

The direct sales effort will be based on a combination of the required frequency of calls and the information required by clients of the company or practice in question. In the construction industry, direct sales/personal calling is also a key feature of brand building and of building and maintaining positive and enduring good-quality channels of communication.

Targeted direct marketing

If this is to be effective, it again requires knowing who the people with influence are. Promotional material, brochures and achievements can then be arranged and presented – and, if necessary, specially prepared – on a client or sector specific basis.

Public relations

Public relations activities exist to generate and support a distinctive and positive organisation or practice identity; to support the direct sales and direct marketing efforts; and to provide a distinctive and positive position that can be supported by other forms of promotion, especially general advertising. It is also essential that the public relations function has a capacity for dealing positively with the press, TV, radio and other media. This in turn means developing links, contacts and networks so that 'good' stories – of creation, achievement and excellence – receive wide and favourable coverage. It is also essential to have the capability to recognise where problems may occur and to possess facilities and expertise (and organisational support) for providing positive responses. As we have seen above, these facilities are essential when dealing with controversial or high profile adverse coverage; and there is a very high propensity for this to exist (or potentially exist) in the industry.

Secondary promotion

Secondary promotion efforts include:

- **General advertising:** especially in the trade press to ensure continuing general awareness. From time to time some companies – especially construction companies – have engaged in television and radio advertising, but this has tended to be during property booms (such television and radio advertising as existed in the mid 1990s is now mostly found in hook-ups with estate agents among house building companies, and out-of-town industrial estate and facilities contractors in areas designated for regional development). Many organisations also include leaflet inserts in the trade press as part of their advertising efforts (see Summary Box 7.6).
- **Logo design and reinforcement:** often found on billboards, hoardings and also on the perimeter fences and at the gateways to project sites. As well as the main contractor, this is likely to include any specialist subcontractors, architects, quantity surveyors and consulting engineers.
- **Attendance at trade, professional and regional fairs and exhibitions.**
- **Sponsorship of events, both professional and social:** professional events include university and college design competitions; supporting university and college tenured posts; and support for research projects. Social events include the sponsorship of concerts, charity events, and local, regional and national fund-raising activities.
- **Attention to general operational features and activities:** this includes matters referred to in the introduction, such as attention to

SUMMARY BOX 7.6 **Advertising**

Construction advertising is targeted at consumers where the consumer is the direct purchaser. This has a special reference to housing extensions, home improvements and refurbishment. Double glazing and conservatory construction companies invest heavily in consumer marketing – especially newspaper and television advertising and leaflets. Architects specialising in home improvements and extensions advertise extensively in the local press. Speculative housing developments are promoted through estate agencies and local media; some also use the wider press – for example, there are regular advertisements for domestic housing in the *London Evening Standard*.

In the 1970s and 1980s, Barrett Homes Ltd undertook an extensive television campaign using the slogan 'Building houses to make homes in'. This was conducted in support of Barrett's extensive house building programme at a time when there was a major 'consumer' drive to become home owners on the part of large sections of the population that had never hitherto had this opportunity. Advertising in this way ceased when the boom ceased.

the way in which the telephone is answered and systems for handling messages. It also includes reference to the appearance, demeanour, willingness and co-operativeness of all staff; the appearance and ambience of sites; the presentation and packaging of company information, designs and schedules; attention to organisation – distinctive features such as site safety; and, again, use of hoardings and boards at the side and entrance to sites.

Other secondary efforts include the following:

- **Estimates and quotations also carry a promotional angle,** especially when tied into deadlines and work content. These elements each reinforce image building, confidence and expectations. Moreover, the price charged itself reinforces expectations and normally has a promotional angle. For example, high price levels are often present as the result of the work needing to be carried out quickly or urgently, or to a premium quality level; while lower prices may not include any after-sales activities.
- **Tendering is also a form of promotion:** It states the capabilities of the company wishing to carry out the work; for a company new to the field, it represents an ambition to work in the field. This is clearly most effective when it is tied precisely into client requirements and matches expertise with demand. More generally, it is often used as a form of promotion both within the industry (so that the presence of the company as a tenderer and potential contractor for particular types of work is recognised); and also among potential client bases (so that the company builds a presence and familiarity in these areas).

- **Speculative designs and proposals are produced as a form of promotion** with a view to:
  - generating work as the result of the specific proposal;
  - creating interest in the potential of a particular area – often urban or dockland regeneration – which may then be developed further once this is understood;
  - creating interest in the capability and expertise of the speculative designer;
  - creating general interest in the design practice as a creative and dynamic entity.

Note

Secondary effort is only useful in relation to the primary or priority. In all construction disciplines, except at the extreme of the micro end, advertising and sponsorship do not generate work. Work is generated through the effectiveness of the primary effort, targeted sales and direct marketing; general confidence and identity can only be supported and enhanced by the secondary activities if they have been generated in the first place.

Tertiary promotion

Tertiary and other more general promotion efforts include the following:

- **Making the most of positive media coverage when it occurs.** This comes in various forms. For example, it is often the case that contractors will place a hoarding at the side of a finished building or facility stating 'Completed six months ahead of schedule' and so on. Companies also make features out of pioneering technology or innovation. For example, Eurotunnel made a feature of its tunnel boring machines at the time of the inception and subsequent tunnelling work on the Channel Tunnel; and Balfour Beatty invited the world's press and television stations along to witness its completion of the A38 river crossing at Plymouth. Effective general support of this kind depends on recognising the opportunities for coverage that exist and having access to those who can make the most of them.
- **Other forms of general positive public relations activities.** This normally means more general work underpinning the determination to be good corporate/community citizens. Current activities that come under this heading at present include general support for schools, colleges and universities (in addition to direct sponsorship, as mentioned earlier); and contributions of general (and not necessarily directly construction related) amenities to communities – for example,

the provision of bottlebank and other refuse collection and waste disposal activities (such as, for example, carried out in various towns in the south-east of the UK by, among others, Costain and Balfour Beatty/BICC).

- **The capability to counter potentially adverse media coverage before it becomes actual adverse media coverage.** This means having trained staff ready and prepared to deal with any enquiry that is made. This, in turn, is founded on a corporate determination that any issue raised will be dealt with in a positive and public manner. Quite apart from the generation of a specific positive company profile, inability or reluctance to deal with enquiries causes the media to search for other information off their back; and when matters are reported in this way, there is a generally negative effect on public confidence and support.

SUMMARY BOX 7.7 General Promotion: Traffic Management

Both construction companies and public authorities have from time to time tried to address the problem of the inconvenience caused by activities that disrupt traffic.

On the political side, the Department of the Environment created the 'cones hotline'. This encouraged motorists who considered themselves disrupted for no good cause by traffic management schemes to complain to the Department with a view to seeking redress, or else getting the works inspected with a view to making them less inconvenient for the public at large.

Urban one-way systems are designed with a view to keeping traffic flowing as quickly as is safely possible in large concentrations of the population. They fall into disrepute where this is not perceived to happen. They also fall into disrepute where there is perceived to be no sympathy on the part of the traffic planners for the needs and wants of the public at large.

According to figures released by the Department of the Environment and the Association of Local Authorities in 1996, complaints about either form of traffic management have all but dried up. The inconvenience has become institutionalised into society at large; and there is a large measure of acquiescence (if not direct acceptance) that this form of disruption is an inevitable consequence of the current state and development of society.

CONCLUSIONS

The principles and importance of promotion are critical to the successful operations and activities within the industry. It is also clear that many organisations – including the largest and most prestigious – could do a lot more promotion, while others could make much more effective use of their existing efforts and budgets.

The aim of all promotional activities is to bring existing or potential customers and clients from a state of relative unawareness of organisational activities, capability and expertise to a state of actively seeking them out and adopting them. The spectrum for this may be identified as follows:

Unawareness	Awareness	Interest/ 'Would you use us?'	Desire: `Will you use us?'	Conviction	Adoption/ purchase/ commission

Some of this reluctance to promote and to engage in marketing activities clearly stems from a fundamental lack of understanding of what promotion is and how the tools of the trade can best be put to use in all areas of the construction industry. It also arises from a traditional lack of effective promotion of activities in the industry as a whole, though some companies (especially the small to medium sized) do promote themselves very successfully. Others – especially the largest companies – have tended to rely far too heavily on corporate name, reputation and image alone. They have failed to recognise the opportunities available, the advantage to be gained – and the business to be gained. This refers especially to direct sales and public relations activities – in dealings with the media, potential and actual clients.

It is necessary to recognise that much of the incursion from overseas companies into the UK domestic industry has come about as the result of effective promotion. The best global companies not only produce their distinctive expertise and competitive advantage, they also heavily promote it – to the extent that clients and potential clients are prepared to give them a chance at the expense of the domestic provider. This arises from applying effectively the lessons and principles indicated – and, above all, the importance placed on face-to-face and direct contact – and this is the precursor to all effective market entry and development.

At the heart of all effective promotion is empathy and the use of language, the ability to communicate with actual and potential customers and clients so that they understand in their own terms what construction sector companies are offering them and the advantages of conducting business with them.

Successful construction promotion therefore places its emphasis on the use of presentations, direct communications and company literature – often specifically produced for individual clients – rather than more standard forms of consumer advertising. This emphasis is based on confidence and certainty in capability and expertise. It must also be capable of taking and understanding the wider view and, where necessary, including some faculty for handling those matters that must be presented to the company's best advantage even if they are not totally within its control.

Effective promotion depends on full involvement of high-quality and trained staff. They must be capable and confident of dealing with the media, public groups, vested interests and other lobbies; and politicians and the governors of public service facilities and utilities. Direct sales and marketing staff take time to get to know their clients and potential clients and the specific demands of each – and, again, undertake to produce specific information in the forms required. They also build effective and continuing liaison, in effect carrying out the groundwork for effective and successful tendering and contract agreement activities.

Successful promotion depends on recognising the importance and value of the activities involved and accepting them as an integral part of the investment necessary in the pursuit of successful construction activities. It depends on recognising the interdependency of all of the different professions involved and that where the efforts of one falls short, it affects the industry as a whole – and also specific operations – adversely.

Different clients and client groups have different expectations of contractors, architects, designers, planners and consultants; and this reinforces the need to produce specific material for specific sectors.

Finally, attention is drawn to the examples indicated. The purpose of these is to emphasise the benefits that accrue from the effective use of the right promotional activities and forms in construction; the potential that is available to companies that do not already engage in taking them up; and, whatever the media selected, that effective promotion is dependent on the presentation of benefits, and the generation of the positive aspects of confidence and certainty that is one of the cornerstones of all effective marketing.

8 Information

The following knowledge is essential:

- **Total market volume and value:** the number of actual clients, the number of potential clients; the numbers who use you as distinct from others; the numbers who use others rather than you; the income availability in the total market; the income gained by you from the market; the potential income available to you from the market.
- **Market nature:** the extent to which the client base is fixed or fluid; whether the market is expanding, static, stagnant or contracting; the duration of the market; certain, likely and possible changes in the market – especially in terms of client confidence and expectations; other possible and likely trends and developments.
- **Market locations:** and the key pressures that these bring, which means understanding the cultural pressures and boundaries (see Summary Box 8.1), and especially those constraints that cannot be altered – such as religious pressures and tendering methods in overseas and other non-domestic markets. It is no use having excellent construction, design or consulting expertise if it cannot be made acceptable to the target market.

SUMMARY BOX 8.1 — **Cultural Boundaries**

The cultural boundaries have to be understood in every context and situation in which marketing activity is envisaged. A critical part of effective marketing means having the capability and willingness to work within these confines as they apply to the client base and to potential client bases; and also recognising the pressures that they bring to bear on the ways in which work is carried out by the contracting organisation.

A simple definition of culture is 'the ways in which things are done here'.

The main pressures and boundaries that have to be learned and understood if marketing is to be successful are:

- **History and tradition:** the origins of the client organisation and client bases; strong social and historical pressures; philosophy and values; the ways in which these have developed.
- **Nature of activities:** both historical and traditional, and also those current and envisaged; this includes reference to the general state of success and effectiveness.
- **Technology:** the relationship between technology and organisation

structure; levels of technology, stability and change; levels of expertise, stability and change.

- **Past, present and future:** the importance of the past in relation to current and proposed activities; special pressures (especially struggles and glories) of the past; the extent to which the organisation (both contractor and client) is living in the past, present or future; the pressures and constraints that are brought about as the result.
- **Purposes, priorities and attention:** in relation to performance, staff, customers, the community and environment; to progress and development; the purpose, priority and attention under which marketing activities are engaged.
- **Size:** and also the degrees of formalisation of structure that this brings. This is particularly important when getting to points of authority within client organisations. It is also particularly important when trying to get the say-so for marketing activities within contracting and supplying organisations.
- **Location:** geographical location, the constraints and opportunities afforded through choosing to be in particular markets; and the specific, political, economic, social, legal and environmental constraints that this brings. It is also likely to include recognising and understanding prevailing local, national and regional traditions and values (in some cases, this means understanding ethical and religious pressures).
- **Management style:** the extent to which this is open, informal, closed, formal, regularised and proceduralised. It also means recognising in contractor–client relationships the stance required by contractors that stands the best possible chance of engaging a relationship that is likely to lead to work. Given the critical importance of personal and professional understanding in the effective marketing of construction industry activities, such knowledge and understanding are vital.

- **Market expectations:** this means recognising and understanding what the market expects as the result of commissioning construction industry activities – and matching and presenting expertise in ways that reflect this. It also means having a clear understanding of the performance and usage of the completed facility in terms of the expectations of the future and long-term users if these are different from the client.
- **Competitor knowledge and understanding:** full competitor analysis (see Chapter 8) extends total market knowledge and understanding as follows:
  - competitor strengths and weaknesses relative to yours;
  - *competitor* advantages, the extent and nature of the work that these bring in, compared with *your* competitive advantages, and the extent and nature of the work that these bring in;
  - other reasons why the client bases use competitors in preference to you, and you in preference to them;

- competitor size, expertise, market share, market potential and reputation;
- competitor marketing methods and their overall relative and perceived effectiveness and success;
- competitor reputation and the reasons why this is held.

From this it is possible to determine the extent to which market share may be taken from competitors, the marketing activities necessary to do this and the length of time that this is likely to take.

- **The product service mix required:** at the one extreme, this refers to the physical quality and attributes of the completed facility; at the other, clients commission work on the basis that as there is likely to be no real difference between the product whoever carries it out, the service level and support also provided may be critical factors.
- **Client knowledge and understanding:** knowing their needs and wants as follows:

  - whether the contractor–client relationship is to be long-term; whether it has the potential for being long-term; or whether it is certain to be a one-off only (and whether there are possibilities of developing this once work is engaged);
  - why they have turned to you, why they have commissioned work, especially politicians and other shapers of public awareness and opinions;
  - chance remarks, which may be worth following up and which may give some form of general information about particular activities;
  - pavement questionnaires, which, however unscientific, may give particular pieces of information that may be worth following up.

These all clearly have their place. The key to using them effectively is to remember the context in which the information was gained and what was actually said (as distinct from what could, should or might be inferred). It may also be important to consider whether what has been learned from these sources accords or contradicts the volume and quality of other information already available; and especially where a contradiction has occurred, to decide whether or not this new point of view needs further pursuit.

MARKET RESEARCH

Market research is concerned with the collection, reporting, analysis and presentation of data for the purposes of market awareness, analysis, understanding and decision-making. These purposes range across the spectrum from the needs for general information gathered from a wide or

LOW COST/ UNCONSIDERED	MEDIUM COST	HIGH COST/HIGHLY CONSIDERED
• Product testing • Awareness • Colour, logo recognition • Other senses: taste, smell, touch	• Product testing • Additional feature awareness • Satisfaction surveys • Attitude surveys	• Extensive client liaison • Personal and professional direct relationships

Figure 8.1 Market research: basic approaches.

very general point of view on the one hand, to specific, detailed and in-depth information on a particular issue on the other (see Figure 8.1).

Genuine research (as distinct from information awareness) is always carried out in response to a given brief or with a stated purpose (however general that may be). A methodology – means of investigation or research – is chosen that is reliable and valid. Within any prestated constraints or imperfections, it must be capable of sustained, rigorous and, often, sceptical questioning and debate (see Summary Box 8.2).

SUMMARY BOX 8.2 **Research Imperfections**

Given the imperfections of marketing information available anyway, the need for a genuinely rigorous approach when pursuing specific and defined research is clearly essential. As examples of how supposedly overwhelming data are corrupted and therefore invalid and unreliable:

- Lobbies for and against particular developments may gain support, for example through shows of hands at public meetings and signatures for petitions; through group and peer pressure (e.g. individuals may give their support to a particular point of view whether or not they believe in it or indeed have any strong feelings on the matter, because others around them do); those who conduct street surveys are influenced by the appearance and demeanour of those who may approach them, and the response is influenced by the demeanour and approach of the surveyor.
- Lobbies for and against particular developments may gain the support of a VIP whose influence may have little or nothing to do with knowledge of the matter in hand. As examples and leaving aside the particular merits and demerits – in 1986 a bypass scheme in Gloucestershire was killed off by the intervention of one local resident, the then Secretary of State at the Department of the Environment, Nicholas Ridley; in 1987 the integrity of the extension to the National Gallery was destroyed for the foreseeable future when the Prince of Wales called it *'a monstrous carbuncle on the face of a much loved old friend'*.

Any opinion is therefore coloured by these approaches and imperfections. It is therefore essential to recognise that they exist. Research can then be structured to accommodate their effects rather than compounding them.

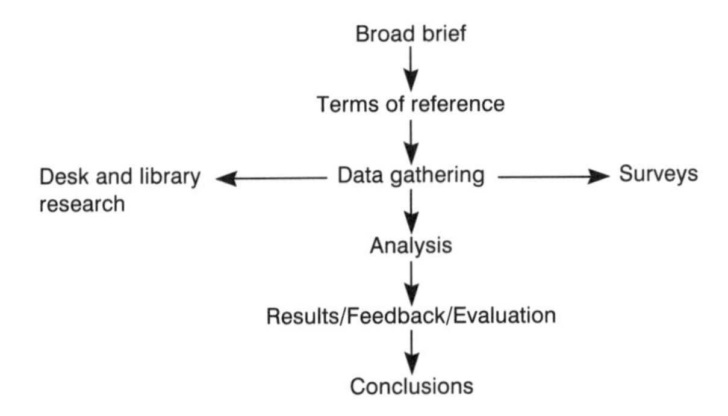

Figure 8.2 Market research methodology.

The normal research methodology used is:

- Defining the area of investigations in precise terms – and this applies, however precise or general the particular area is to be. The final outcome of this is the brief or terms of reference to which the researcher will adhere.
- Developing the research plan: which consists of (a) identifying the scale and scope of secondary and desk research; and (b) identifying any primary research required and the means by which this is to be carried out.
- Whatever methods are to be used must be appropriate to the situation and be capable of fitting into any specific constraints – especially time and resource pressures.
- Gathering and analysing the information: using the methods chosen in the ways agreed and drawing conclusions that are supported by the information.
- Presenting the findings, conclusions and recommendations to those who commissioned the research.
- Reviewing, monitoring and evaluating progress at all stages.
- Identifying areas of imperfection where little or no useful information can be gathered and where further work is required.

The process is summarised in Figure 8.2.

Clearly, there are distinctive consequences, implications and obligations to be understood when undertaking work from this point of view:

- Why they have turned to you, why they have commissioned work or are interested in commissioning work; their knowledge, understanding and perception of your expertise.
- Their key pressures and constraints – time; cost/price; facility, longevity, durability, flexibility, appearance, materials used.
- Where the particular commission lies in the client's priority order.
- Their non-professional pressures – political, social, ethical, religious.

SUMMARY BOX 8.3 **Use of Market Research Agencies**

These are best used when they have distinctive expertise in the areas of market research required and when the particular organisation does not have its own capacity or resources. The following types of agency are available:

- **Sector specialists**: experts, for example in civil engineering, house building, building products and design, who can be commissioned to seek out and research new markets for existing products and services.
- **Regional specialists**: experts in the population, industry, commerce and public service composition of particular localities who provide specific information about a targeted or defined sector (e.g. number of private houses without double glazing, numbers of council houses without double glazing, the names and nature of those who commission civil engineering projects).
- **Construction industry specialists:** who concentrate their energies on market knowledge rather than designing, planning and building things.
- **Consumer and client analysts:** who define the spending priorities and habits of the given client group, their priority orders and their propensity to spend on the particular product, project or service.
- **Economic think-tanks and forecasters:** who provide detailed information and predictions for the future at the macro end and who are of special value to large architecture, design, building and civil engineering companies.
- **Service research firms:** who gather consumer and trade information which they retain on databases or present as reports and which they sell for fees to clients.
- **Research-led firms:** who are hired to carry out specific research projects; their expertise is in research methods rather than the industry or commercial sector.

Whichever is required or chosen, the critical element is agreement over the brief and methodology to be used. Terms of reference may be limited or skewed by the agency's perception of what is required; or the agency may find itself having to fill in the gaps in the brief to the best of its ability. The brief is always bounded by cost and time, and the balance between having good information at a given cost and excellent information at a higher cost has always to be struck.

- It is also necessary to understand the nature of the client's decision-makers – whether they are construction experts, accountants, politicians, governors, shareholders' representatives – and to be able, therefore, to present your expertise in ways understandable and acceptable to them.
- Market profitability – the extent to which the given market is going to provide a basis for successful business; where the work is long-term, providing steady but relatively assured income levels; whether

it has the potential for providing high income levels; the extent of investment required to secure the desired/available level of profit; or whether the present work and that for the immediate future are themselves an investment (i.e. carried out at a loss or reduced margin), the outcome of which is intended to be securing a long-term, ultimately profitable, working relationship and pattern in the sector.

- Potential for marketing activity – this refers to:
  - (a) the extent to which the purpose of marketing activity is to take work from competitors, to extend the market and client base, or a combination of the two;
  - (b) the extent to which doing one job for a particular client may lead to additional work alongside that in hand at present;
  - (c) the extent to which satisfactory completion of one piece of work is genuinely likely to lead to further commissions, either from the particular client or from others in the sector.

Marketing information and research

The basis of all this is excellent and effective information, essential for the successful management of all aspects and functions of all organisations, and marketing is no different (see Summary Box 8.4). Marketing information is required in order to gain as much knowledge and understanding as possible about:

- the client bases, markets and segments served;
- the effectiveness of organisational marketing strategies, approaches and activities;
- the activities and effectiveness of competitors;
- specific factors such as sales figures, effectiveness of marketing campaigns and initiatives.

The outputs of this information, if it is used effectively are:

- risk reduction, as a result of a greater understanding of the prospects and potential of particular activities;
- the ability to analyse products, services, the state of the market, client base, client confidence and client propensity to spend;
- the ability to analyse accurately product and service mixes and portfolios;
- support for accurate and effective decision-making;
- the implementation of successful control and progress monitoring activities when these are set in motion;
- the availability of a comprehensive body of both detailed and general information as an aid to continuous development of expertise, professional knowledge, understanding and awareness.

SUMMARY BOX 8.4 **Marketing and Information**

The best consumer organisations gather minute details about their product, consumers and users. The following are examples of global, sectoral and local marketing information:

- One million Americans drink Coca Cola with their breakfast everyday. One billion people buy Coca Cola in the world everyday.
- Kleenex discovered that the average person blows his or her nose 256 times per year.
- The highest *per capita* consumption of cigarettes in the European Union is in Austria.
- In Moscow, McDonalds is perceived as a luxury restaurant rather than as a fast food outlet; 82% of the local customers go there on special occasions only.
- The Black Lion sports centre in Gillingham, Kent, gathers information on the location, spending habits, family size, home and car ownership, white goods ownership and propensity to spend on holidays of all their clients. From this it relates facility usage as follows: regularity and frequency; priority; factors affecting usage of the sports centre; factors affecting propensity to spend elsewhere; pricing possibilities and opportunities; opening and facility availability usage times.

In each example, the common factor is knowledge. If a million consumers are drinking a particular product with their breakfast everyday, then this has implications for both continued marketing of that product and also the alternatives offered by competitors. In order to support an 'activity' that is carried out 256 times a year, the product has to be priced accordingly. The presentation of something that is regarded as a luxury in one part of the world is going to be very different from presenting it as ordinary, common place, fun (or whatever) elsewhere.

From the other angle, the gathering of this detailed information enables, for example, the sports centre, to target accurately: frequency of visits; main users of particular facilities; price ranges and income possibilities; and regular, frequent and profitable users can be targeted in terms of consulting on service upgrades and improvements. Infrequent and overtly unprofitable users can be targeted with a view to establishing the extent to which it is possible to turn them into regular and profitable users, and what activities the centre would have to engage in to make this possible.

Costs and benefits of information

No information is ever perfect. However comprehensive its gathering, full information is never available. The problem therefore lies in deciding the balance of volume, quality and analysis, and this is one of the critical judgments of any marketing manager. Computerised and electronic information systems have come to be the industry standard and this has led to great increases in volumes of information available both to companies and individuals.

Quality

The quality of information is much harder to define. It is a combination of:

- the reasons for which it was gathered and whether these were the right reasons or a fully comprehensive set of reasons;
- the questions that were asked, and whether these were the right questions and a fully comprehensive set of questions;
- the costs incurred in gathering the information – the time allowed in which to gather it and whether this was adequate;
- the uses to which it is to be put;
- the conclusions, forecasts and extrapolation methods used;
- the personal and professional expertise of those gathering it and those using it;
- the choices apparent as the result of having the information available and the opportunities and consequences afforded;
- where the imperfections lie, where the areas of greatest risk are, and whether there is any need for further information to be gathered.

Balance is therefore clearly required. It is possible to become trapped into doing nothing but gathering information and this clearly leads to corporate inertia and paralysis. It is also used as a device in organisation *realpolitik* systems to kill initiatives. On the other hand, progress on the flimsiest knowledge is highly risky and (with very few, usually highly publicised, exceptions) normally results in failure. This is especially critical when dealing with limited numbers of high-value customers and clients, as is the case with the construction industry.

Analysis

This is again imprecise and proper analysis depends entirely on the appropriateness of the data and the expertise of the analyst or user. It is a combination of:

- sectoral knowledge;
- ability to use any statistical, financial and other qualitative and quantitative methods required, as appropriate;
- the soundness of judgment and evaluation;
- the context of analysis – the best, medium and worst outcomes, acceptability and political factors;
- the ability to recognise any critical factors (see Summary Box 8.5).

SUMMARY BOX 8.5 **A Bridge Too Far: The Three Photos – Critical Factors**

'A Bridge Too Far' is the name of the film made by Richard Attenborough about the war-time Arnhem landings in 1944. There was great political and military pressure to carry out the particular attack – which consisted of dropping an army of 30,000 behind the German lines in The Netherlands, in 1944. The stated purpose was to increase the Allied advance and 'shorten the war'.

The plan was devised on the basis that 'it was a good idea'. All the information gathered was analysed on the basis that the initiative would succeed – not that it might not. That the Germans were in retreat was taken as an absolute fact; it was not researched.

The week before the landings were due to take place the reconnaissance mission of the area produced three photographs that showed beyond doubt that a large and well-equipped German army was undertaking rest and recuperation in the precise areas where the paratroops were to be flown in.

This information was discounted – it was simply edited out by the perceptual processes of those in charge. The evidence therefore was that the initiative could not and would not succeed. Those in charge nevertheless went ahead regardless – and the initiative failed.

In this context, the points to be made are as follows:

- that information is used in the broader context of organisational and operational drives and activities;
- that the best, medium and worst outcomes of anything envisaged are apparent at the outset;
- that the likelihood of problems occurring and their nature can be clearly foreseen at the outset of any activity;
- that once it becomes apparent that there is a high risk of failure, this does not mean that activities do not go ahead;
- that the 'what if?' approach to activities must at least take into consideration the consequences of failure or shortfall in absolute success.

The desired outcome is therefore as full an understanding as possible of the opportunities and pitfalls available and apparent, and identification of the critical areas for success. If any of these factors have not become apparent then further work is necessary; while if such matters are now apparent, a judgment is then necessary as to whether further work must be carried out and if so, what its purpose should be.

Sources of data

Data can be categorised as primary or secondary:

- **Primary data**: obtained by organisations and individuals directly through observation, surveys, interviews and samples, and using methods and instruments drawn up specifically for the stated purpose.
- **Secondary data**: from other data sources, such as official statistics provided by government sources and sectoral data gathered by employers' associations, federations and marketing organisations.

Uses of data gathered

The use of secondary data always involves taking information that others have gathered, and interpreting, analysing and using it for purposes different from those that the original gatherers designed or intended. There may also be variations in definition or coverage that have to be taken into account.

The decision to gather primary data or to use other sources will depend on:

- the nature of the information required;
- its availability from sources other than primary;
- its range and coverage;
- the field of enquiry and its size and scope;
- the accuracy of the data required;
- the date or deadline by which the information is required.

It depends also on the purpose and aims of the enquiry to be made and the uses to which the information is to be put.

The data thus gathered are then classified into groupings or classes with a common element for the purpose of analysis, comparison and evaluation. From a marketing point of view, the main purpose of this gathering and assessment is to provide background information that is accurate and quantifiable. This, in turn, becomes the basis for accurate planning, forecasting, projected activities and decision-making; and at a strategic level, to provide support for the accuracy of general direction.

It is useful, therefore, to identify the different sources of statistics, information and knowledge that are available and their particular uses. These are:

- **Government statistics**: available from government sources and useful as general indicators of the state of business and economic activity and confidence in the particular sectors under consideration.
- **Sectoral statistics**: produced by trade federations, employees' associations, employers' associations and professional bodies for the support

and enlightenment of members of organisations, and to contribute their knowledge and awareness of the global aspects and overview of their own sectors.

- **Market research organisations**: which hold data on vast ranges of issues that they promulgate and sell on a commercial basis to those requiring it; and which they may gather on a contracted basis. The main initial value of this is to indicate the general state of business and range of business opportunities that may be available. This information also may be used as a prelude to conducting and commissioning more specific investigations.
- **Local government bodies**: which hold a wide range of general data on the composition, social state, occupational range and population structure of those who live in particular parts of the UK. This is published by local government and municipal departments and, again, is a useful precursor to more rigorous investigation. It also enables initial general assessments, knowledge and understanding to be gathered by organisations when considering moves into new areas.
- **Public inquiries and investigations**: which generate a great amount of information concerning particular initiatives (e.g. on urban development, infrastructure projects) that are often a useful initial point of reference for those planning to go into similar ventures in the future.
- **Organisational statistics**: which are gathered internally for specific purposes. Organisational statistics may be available in sufficient detail and quality to preclude the necessity for primary investigations. In particular, good organisations have the following marketing information available and ready to hand – market assessment information; general means of gathering marketing information (e.g. through sales teams); effectiveness of the targeting of marketing activities; effectiveness of general marketing activities; effectiveness of the total marketing position (see Summary Box 8.6).

A question of balance

While primary data are gathered for given, direct and stated purposes, it is also much more expensive. Expenditure on primary search has therefore to be cost effective, in the knowledge and understanding that whatever is required is either not available elsewhere or not available in the form required.

SUMMARY BOX 8.6 **Marketing Information**

Specifically, the following marketing information is essential:

- **Market targeting**: information available for using and evaluating the effects of marketing campaigns; effectiveness of the targeting of marketing activities; agreed means by which this effectiveness is to be assessed; gaps in information; inappropriateness/appropriateness of information; steps that need to be taken to remedy any shortfall.
- **Sales**: by product; by product cluster; total range of products; by outlet; by location; volume and quality; demands for returns; after-sales demands; the number of times that guarantees are invoked; complaints; sales blockages.
- **Production**: deliveries; product output volumes; product to market time; other time factors; quality factors; volume factors; number of complaints per site, factory, batch, unit, production run, location, project, product; production blockages; supplier factors; distributional factors.
- **Financial**: total costs; cost breakdowns – by site, division, department, function, location, occupation; fixed costs; variable costs and causes of variability; marginal costs; budget and budgeting processes.
- **Performance measures**: those chosen; why they were chosen; who chose them; their appropriateness or otherwise; the context and expertise used to measure them.

From a marketing point of view, it is clear that there is a vast range of information available for use. Overall, the effectiveness of this use depends on the organisation's ability to gather and store this information, on the capabilities of those involved to identify what they want, when they want it and how it is to be evaluated, and the priority that the organisation puts on these activities.

Primary data

The main methods available for the gathering of primary data are as follows:

- **Observations**: which are subject to perception and interpretation; the use of the senses; and the ability to take in enough of a particular situation to form a sufficient understanding of which judgments can be made. In some cases, observations are also limited by the ability of those involved to define hypotheses and devise means for testing these in ways that are capable of being validated.
- **Case histories and examples**: which are again subject to perception and judgment, except where cause and effects can be directly related (for example, where £10 was made because ten items were sold at £1 each).

- **Document analysis**: subject to the knowledge, quality and judgment of those who originally produced the documents, as well as being subject to the interpretation of those currently using them.
- **Questionnaires**: again limited by the capabilities of researchers to define their purpose, ask the right questions, interpret and analyse the responses, and draw conclusions from the information of material gained. Limitations are also produced by situational factors, individual priorities and perspectives, and time constraints. These are continuously influenced by their environment and can quite legitimately provide a different set of answers to the same questions within moments if circumstances suddenly changed. This also applies to structured interviews and the media used – whether face-to-face or telephone; whether it is an individual or group situation; time constraints; the attitude of the questioner; the personality of the questioner; the importance of the subject matter to interviewer and respondent; the extent of mutual respect; speech patterns and emphases on different words.

  Responses are also conditioned and limited through the responders not knowing the answers to questions or only knowing a part; they may also tell the interviewer what they think he or she wants to hear, or what they think the answer should be; they may lie; they may give no answer; they may give an answer at variance with their own views or understanding because they perceive that this is what is expected of them; or they may just make something up. Moreover, responses are conditioned by wider situational factors and constraints – such as matters of confidentiality; the use to which the information is going to be put; the use to which the respondent perceives that the information is going to be put; and any opportunities or threats that are known or perceived to arise as a result of giving particular answers (see Figure 8.3).

Accuracy

The accuracy of any data depends on the way in which it is gathered, the quality of the actual data gathering and any rounding at the end of it. If a survey took a sample rather than dealing with everyone or everything concerned in a particular activity, the results may indicate particular conclusions very strongly, but these will only be proven if the entire sector is surveyed. If there is a flaw in the statistical methods used or if the wrong questions are asked, the results will also be flawed and inaccurate. Rounding of numbers is widely used and has also to be seen in context and as a limitation – for example, market volumes are given very often to the nearest £100,000 or the nearest 1000 products.

Social survey and market research organisations consequently go to a lot of trouble to make their surveys both valid and reliable through the

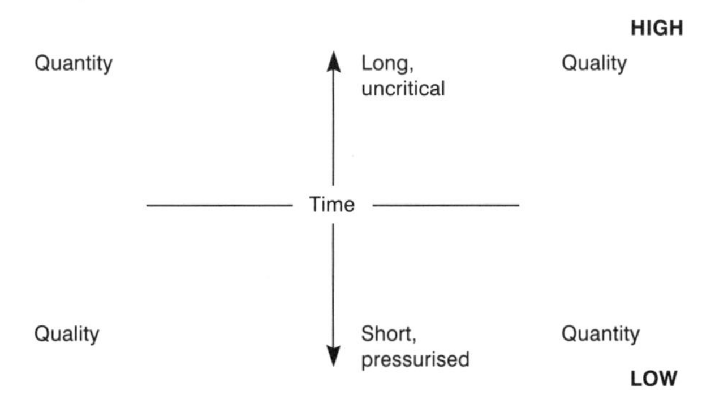

Figure 8.3 Key features in the gathering and analysis of primary information.

establishment of proper objectives; the accuracy of the commission; the design of questionnaires and other survey instruments; the provision of high-quality and rigorous training of those who are to conduct the surveys; and the recognition and promulgation of limitations on the research to be carried out. Any findings, analysis and conclusions are thus seen in the broader context.

Other elements that impinge on the accuracy of data gathered and availability are to do with: time lapses between the gathering, promulgation and interpretation of data; sampling errors; analytical errors; and inexplicable inconsistencies. These are almost inevitably compounded where extrapolations and forecasts are required; and the greater the time pressure or the smaller the sample, the greater the likelihood of this occurring.

Those involved in marketing must therefore strike a balance between making the best use of the information that is available and developing a positive and healthy scepticism, questioning technique and enquiry into any data presented to them. This is not to promote inertia or the continuous circle of information gathering. Those involved should ensure, however, that they do question and consider all aspects and the full implications of statistics as the basis for accurate decision-making and initiative formulation.

Internal information

Internal information may be primary – gathered by researchers from internal sources for their own purposes; or secondary – gathered from existing sources and databases elsewhere in the organisation. For those concerned directly with marketing, the most valuable internal information concerns production and sales. This is required as follows:

- **Overall levels of production**: background product knowledge in order to be able to give an informed answer to clients, especially when urgent, high-volume and high-quality work is being requested.

- **Specific levels of production**: to be held in response to specific client requests.
- **Production costs**: so that accurate, effective and profitable estimates can be made and which indicate (though they do not prove) best, medium and worst levels of sales, and levels below which a contract may become unviable.
- **Management accounting information**: so that the ways in which the organisation considers and calculates its cost and charge bases may be understood (even if not always agreed with).
- **Time from order to production**: which identifies problem areas, areas for improvement; and which may be a critical factor in dealings with clients.
- **Time from order to completion**: which again may be critical in dealings with clients, especially in relation to important and urgent requests or as part of getting on to a preferred contractor list.
- **Time from contact to response**: often of critical value as the first step in establishing a profitable contractor–client relationship.
- **Customer profitability**: analysis of contract list and currency by contract income/contract expenditure; and this may be broken down further into contract components, total business volumes, individual contract values and total relative income.
- **Product profitability**: part of value analysis; and used especially so that the individual and total contribution of each product, project and service can be established. It also identifies product mixes under different headings, especially those of the Boston Matrix and Drucker Matrix (see Chapter 5, page 118).
- **Budgeting**: as part of analysing the adequacy and effectiveness (or otherwise) of budgeting processes and allocations.
- **Specific information**: internal information systems must also be capable of providing specific answers to specific questions. For example – the number of sales calls per genuine prospect; the number of genuine prospects per order; the number of sales in the given region or time period; average income per order – in total, by sector, region, time frame, activity – all of which can then be used for assessing the effectiveness of marketing activity. This then gives detailed background information and the context in which the whole marketing activity can be seen.

External information

External information is available from all of the secondary sources indicated above and the volume grows larger by the day. Organisations also commission their own external primary searches and surveys. The key to its effective and successful use lies in recognising and understanding those sources that are of greatest value and broadening the general field of enquiry as far as possible – while at the same time making sure that what does come in continues to be useful.

External information may be classified as:

- **Active**: that which is actively sought by the organisation, whether for specific requirements or general enlightenment.
- **Passive**: that which arrives at the organisation or on the desks of particular individuals and which may nevertheless be useful.

The following external information is essential:

- **Environmental**: as an aid to knowledge and understanding of the markets, sectors, segments, nations, regions and locations in which activities are to be conducted.
- **Markets**: general market information is normally made available through professional associations, trade journals, library and archive facilities of the client groups. General information is also gathered as a part of direct marketing and sales calls, and it is vital that this is assimilated, understood and followed up where necessary or desirable.
- **Competitors**: the purpose of gathering and analysing competitor information is part of the process of understanding the current state of the industry/profession and its structure. More specifically, it provides a general understanding of perceptions and misperceptions and the relative nature of competitive advantage. It gives a clear understanding of why, where and how some organisations succeed and others fail – the difference between an organisation's perceived view of its competitive advantage and those actually regarded as important by the industry itself.

  For example, the cost leader organisation in a sector may perceive its advantage as its ability to compete on price, and the more it fails to get work, the lower it drives its prices. This is only the right course of action if the client base is seeking price advantage. If it is not seeking this, the cost leader must find other points on which it can compete effectively.

  It is also essential to be able to identify early potential entrants from elsewhere (e.g. from different regions or overseas), the consequences and threats (and opportunities) that they may bring and what moves are necessary to keep them out or limit their effect.
- **Clients**: clients should be studied in the same way as competitors. The ideal output here is a genuine understanding of what persuades them to use you as distinct from the competitors, and what persuades them to use competitors rather than you. Such data are essential on both individual clients and also client groupings and bases.

It should also be clear from this how long lasting particular contractor–client relationships are and are likely to be; and the extent of the certainty, likelihood and possibility of these in the future. It is also necessary to look at the softer side of what constitutes client expectations and satisfaction – especially personal and professional relationships – and how these can be developed.

- **Consumers**: the particular problem for all activities in the construction industry is the ability to estimate the impact of projects and initiatives on the eventual users of the particular facility or item. There is therefore a need to understand, at least in general terms: current trends and patterns in consumer behaviour; components of consumer satisfaction; the relative value placed on different construction activities (e.g. house building, refurbishment, double glazing, do-it-yourself); the relative value placed on different public facilities (e.g. sports centres, car parks, shopping precincts, railway stations); and the difference between attitude, acceptance and usage (see Summary Box 8.7).

SUMMARY BOX 8.7 **Consumer Information in the Construction Industry: Examples**

- **Double glazing**
  There is a widespread perception that householders and commercial operators who order double glazing assume that they are being ripped off. Once they have decided that the work needs doing, they nevertheless go ahead with it. There is therefore a fundamental difference between the attitude, which is very negative, and the response, which is very positive. The (perceived) benefit of having the work done outweighs the (perceived) quality of contractor–client relationship.

- **Car parks**
  Urban multistorey car parks are widely perceived as eyesores, environmental and aesthetic disasters and are therefore often controversial at their point of conception and decision to build. During building and completion, because of the very fact that they are normally sited for the convenience of traffic and car drivers, they are often very disruptive to everyday personal and commercial life. Once completed, the attitude often changes completely as people get used to their existence and start to use them for their own convenience.

- **Design**
  Nothing illustrates the passage of getting anything from revolutionary concept through acceptance, normality and antiquity to obsolescence (see Product Life Cycle, Chapter 5, page 118) better than building and facility design. Moreover, the overall perception and attitude changes – from environmental blight, to environmental disruption, to environmental enhancement, through confidence, usage, disuse, ruin, demolition (or archaeology).

In this aspect, attitudes within the design and architecture sectors may be very different from those demanded by clients and consumers; and these may be very different again, between clients and consumers.

- **The built environment**
  This refers to the integration of different, divergent and often controversial initiatives into the wider environment. All urban and industrial developments are affected by this to an extent.
  An example of this is the development permitted on the islands of the city of Venice. There are designated areas for general modern development to the north and on the landward side of the city's railway station. Within the main city, both new developments and refurbishment of existing buildings have to be approved by a variety of committees in order 'to preserve the traditions and environmental richness of the city'. This contrasts utterly with the fact that the historical city was developed largely piecemeal over a period of about 700 years.

Information gathering and analysis activities are certain to produce a range of attitudes and views along the lines indicated. The key lies in understanding that they are normal human responses and that they do change provided that what is proposed is believed in, of value and will make a positive contribution once completed.

Marketing activities and initiatives

This is the hardest area of all in which to gain real and accurate information (as distinct from informed perceptions). There is a great range of information necessary:

- The costs and benefits of different approaches.
- The impact of different approaches on existing and potential sectors and segments.
- The costs and benefits of doing things because others in the sector do them; the costs and benefits of not doing these things or doing things that way in order to differentiate.
- The returns to be gained from reducing, maintaining and increasing marketing activities, and the mix of marketing activities; and this includes reference to general favourable responses and awareness; new business from existing clients; new business from new clients; and the density and frequency with which marketing activities result in orders for work.
- The returns to be gained from reducing, maintaining and increasing price levels; offering different price levels according to sector, location.
- The returns to be gained from reducing, maintaining and increasing existing product ranges, service levels and support services (e.g. after-sales, facilities management, refurbishment).

- Reference to time lags – the periods of time that must elapse (and this establishes the length and nature of marketing investment required) before given levels of marketing activities can be expected to show results.

The core of the answer clearly lies in the knowledge, understanding and expertise of those responsible for addressing the questions. The problems can be minimised by taking time and resources to establish the precise points at which marketing activities are to be targeted and the rationale that lies behind this approach.

The general contribution of other information to marketing

This refers to the effect that other stories about the organisation have on its marketing effectiveness (see Summary Box 8.8). All such stories affect the wider perception in the environment. They may and do also directly affect client and consumer confidence and propensity to use. Moreover, development and explanation of positive stories and direct and positive responses to negative coverage each emphasise overall perceptions of confidence, and exploit what may be rare opportunities for media coverage.

Marketing information awareness

This is an active consideration of the need for information under the different headings indicated. The following is also essential:

- the need for general market intelligence (on general issues and particular developments);
- the awareness that everything can always be improved – and this includes the volume and quality of information, as well as the products and services on offer;
- straws in the wind – the development of early information and early warning systems that ensure that particular organisations are among the first recipients of information about new developments, initiatives and opportunities.

However general or incomplete such information may be, in certain circumstances it is a powerful reinforcement of the contractor–client relationship. In all circumstances it should trigger off the determination to find out as much as possible, as quickly as possible, about whatever is in hand.

Marketing intelligence

This is a proactive approach to the problem of awareness. It consists of investigating the environment, current markets, potential markets, and

SUMMARY BOX 8.8 **Other Stories**

Channel Tunnel
During the period of construction both Eurotunnel (the client) and Transmanche Link (the contractor) handled a variety of environmental, financial and technical problems. Levels of general confidence remained high because of the positive approach of both companies, the upbeat and positive nature of public relations and general marketing activities. The worst marketing problems to the Tunnel during its construction were caused by the one major factor over which the companies had no control – the high speed rail link on the British side that, when completed, would connect the Tunnel with London and the rest of the UK rail network. The problem was neither of the companies making nor of their choosing, yet both found themselves concentrating resources on providing substantial information, media and public relations coverage on handling this issue and a substantial part of the companies' library and information banks had to be given over to this matter in order to prevent their being tarnished by the lack of provision and consequent lack of confidence elsewhere.

Blue Circle
In January 1997, Blue Circle announced a long-term (five-year) pay deal that would give pre-agreed pay rises for the period 1997–2002 in return for guarantees about job security.

The story primarily concerned staff management and industrial relations. Nevertheless, it received widespread coverage in both trade press and national media with the emphasis on the highly ethical position of the company in its concern for its employees. It also provided the opportunity for the approach that 'we are a highly profitable, extremely secure company producing top quality building products' and that therefore its customers and clients could continue to do business with it in the certainty of a long-term mutually successful and profitable relationship.

internal, actual and potential capabilities of the organisation along the following lines:

- **Undirected scan**: general exposure to information with no specific purpose in mind; though what is discovered as the result may lead to very specific enquiries very quickly.
- **Trade press and literature reviews**: with the purpose of concentrating on particular types of information or particular presentations, discussions and analyses. This may be directed at: the contractor's sector of activities and those areas related to those sectors; the client's sectors and those closely related; academic literature and studies; and related trade press and academic literature to see if anything may be learned from elsewhere.

- **Academic and/or commercial project work**: this consists of having a group of people (e.g. outside students, sponsored students, project teams) working to a given brief with the view to producing 'something further' at the end of the work. Clearly this may be – and often is – both general and inconclusive. The benefit is that both the work carried out and also the summary conclusions and recommendations arrived at as the result may indicate (or not) the need for further efforts.
- **Undirected searches**: these consist of pursuing particular ideas to see what information can be gathered about them and/or to see where else they lead. They may, and often do, consist of no more than feeding buzz words into a book index, electronic or CD–ROM library and database facilities, and seeing what is available as the result.
- **Formal searches**: these are deliberate efforts to secure detailed and specific information, and are usually accompanied by the requirement to work to specific plans, procedures and timescales.
- **Client and competitor searches**: these are deliberate efforts to secure as much information as possible of any quality or volume about clients' competitors. The usual approach is to use the competitor or client's name and/or activities as the buzz word. This is then handed over to an expert in database, CD–ROM and library searches. Those involved are given a wide-ranging brief – often supplemented with specific requests and targets. The result is the gaining of a substantial body of both general and targeted information which can then be used or pursued further.
- **Direct relationship contribution**: this is the general contribution to organisational knowledge and information that is made by the salesforce and others (e.g. public relations) who have direct dealings with clients apart from their specific tasks and duties. It is a key contribution, especially to the background, and a major output of high-quality, expert and well-motivated sales and public relations teams. They are in the best position of anyone to pick up 'straws in the wind', general perceptions, market intelligence and other undirected information.

Whatever the research and information carried out and gathered, the important thing is their subsequent use. Used properly, they are an aid to effective decision-making and support for particular courses of action. They provide the means for assessing why successes occurred (all too often the reasons for success are never properly evaluated), and a basis on which the analysis of failure or action that fell short of full success can be analysed.

More generally, summaries and synopses of the information should always be made available to marketing and public relations staff. They should also be available to production and service specialists and direct practitioners – anyone, in fact, who needs to deal with clients and potential clients on an informed basis.

SUMMARY BOX 8.9 Other Sources of External Information

The contribution of all sources should be acknowledged; and the potential that all sources bring, however imperfect, should always be considered and evaluated. Useful examples are:

- **Telephone calls and telesales**: especially important in the sale of domestic and commercial building products and in the marketing of services.
- **Professional networks**: general discussions at the meetings of professional bodies.
- **Professional association gatherings**: which meet regularly on a local, regional, national and professional basis.
- **Market and marketing seminars**: in which one of the perceived direct benefits of attendance is the opportunity to network and talk with peers and compatriots.
- **Enthusiasts**: whether from within the industry or not.
- **Important people**: industry gurus and leaders; politicians; those with extensive influence – especially politicians.

There are therefore distinctive consequences, implications and obligations to be understood when undertaking work from this point of view.

Without adequate information, all marketing activity is conducted in the dark. Success and failure become accidents that arise from carrying out unstructured and uninformed activities. Organisations that do not carry out sufficient research are always at a disadvantage in comparison to those who do. Moreover, the more information available all round, the higher the marketing success rate in terms of calls and tenders that lead to work, and the greater the capability to assess accurately what went wrong with calls and tenders that failed to generate work and with other marketing activities overall. The fact that marketing information can never be fully accurate does not mean that it should not be rigorously gathered, used and analysed whenever and wherever possible. Its inadequacies must be recognised however, and people who take marketing decisions must understand that they do so on the basis of information that is certain to be incomplete.

9 Conclusion: The Future of Marketing in the Construction Industry

The purpose of this final chapter is to identify those matters that are important or are likely to become so across the entire industry in the late 20th Century and beyond.

The industry is in a state of flux. Great change, turbulence and instability are the context in which its activities are set to take place for the foreseeable future. As well as being driven by a technological revolution, the large companies of the industry are now truly global, and smaller companies are being driven to seek work elsewhere than their hitherto accepted spheres of operation.

Other drives for change that affect the construction industry are as follows:

- **Social**: the changing of people's lives from the fundamentals of life expectancy to lifestyle choice; increased expectations of quality of life and quality of working life; increased demands on facilities; increased demands on natural resources; increased demands for physical and information infrastructure.
- **Economic/political**: changes in all government forms; the state of flux of the EU and the adoption of super-national laws and directives; the EU single market; the expansion of the EU; the collapse of the Communist Bloc and the USSR; the emergence of Taiwan, China, Korea, Vietnam and Malaysia as spheres of economic influence; the continuing emergence of Indonesia and the Philippines; the potential emergence of South Africa and South America.
- **Expectational**: the wider recognition that increased quality is available (and therefore expected) on the part of consumers and end-users; specific changes in expectations (especially of quality of finished project and after-sales commitment) on the part of client groups.
- **Legal**: the ability to operate within legal constraints wherever necessary; specific legal constraints that are found in some markets – and within which the industry has to work for the future.
- **Ethical**: taking an enlightened view of the industry's capabilities and commitment; taking a specific view of the organisation's capabilities and commitments and the ways in which these are to be delivered; 'the promise' of project and activity delivery to the highest possible standard.

- **Stakeholders**: a key feature of the marketing activity is 'marketing to different stakeholders'. In particular, this means educating shareholders and other backers to expect and anticipate returns on their investment in terms of long-term commitment, continuing volumes of work, the levels of investment necessary to gain and maintain secure bases for work, and the position of marketing in this. It means accepting responsibility for dealing with legitimate and genuine concerns from pressure groups, lobbies and other vested interests. It also means accepting responsibility for staff – the major concern of all those working in organisations at present (and for the foreseeable future) is job and work security; and this is only to be achieved if skills and qualities are developed in terms of (a) expertise and (b) client requirement.
- **Profit**: 'profit is the by-product of effective activity' – and this means shifting the conception of this from a line on annual reports to a reflection of a state of long termism, and a relative reflection of the permanence of the organisation and its continuity in the pursuit of consumer and client satisfaction.
- **Attitude**: this is the way of thinking about marketing that applies to everybody involved. A complete change is required, away from formalised job positions, systems, order and methods to the realisation that everybody, at every organisation, contributes in their own way towards long-term organisational effectiveness and satisfaction; and that, indeed, this is the only reason for employment.
- **Resources**: the maximisation and optimisation of resources, returns on investment and performance outputs are paramount. Levels of satisfactory performance based on historic activity schemes and the returns that these produce are no longer acceptable or tenable. Neither, increasingly, are they competitive, profitable or effective. There is an ever-growing awareness that resources have been under-used. It is increasingly coming to be realised that attention to these elements is in itself a source of profitable and cost-effective activity. It is also the case that a part of the drive for maximisation and optimisation consists of attention to presentation as well as capability – and to attitude as well as expertise. This is of critical importance wherever client choice is being made on grounds other than capability.
- **Information**: there are competitive and commercial advantages to be gained from the effective and continuous gathering and analysis of market, customer and client information; and of a broader knowledge of the ways in which the industry operates in different localities, regions and countries. The full potential of this is only now being assessed and realised. Sources of information and the speed at which the information can be gathered and processed will greatly expand in the future.

GLOBAL PROJECTS AND ACTIVITIES

There are no barriers to the global operator. There are no areas of activity that are genuinely closed – though restrictions may be placed by quota systems, and restricting rights of entry by particular countries or regions, or by placing legal constraints (such as the need for a local partner) on particular sectors.

Economies of scale, the command of resources and production/output technology, and the need to maximise the returns on these, drive any organisation in this position to seek opportunities wherever in the world they may arise.

Levels of investment also make it very difficult for any organisation that is not of this size or cannot command this level of resource to compete on this scale. However, opportunities abound as subcontractors, local partners, minor partners in joint ventures, and specialists. Smaller activities are also required alongside great infrastructure projects.

So construction projects of the size and scale of the Channel Tunnel, Scheldt Barrage and Hong Kong Airport bring with them requirements for the development and construction of access and infrastructure, housing, commercial buildings and facilities, and industrial estate type projects; and also for leisure facilities, education, health and social provisions; and, therefore, for landscaping and other aspects of urban, regional and rural planning. Many of these projects are politically as well as economically driven – and the opportunities are there for anyone who can present as well as deliver their capability.

Indeed, part of the political drive in many cases (especially in the developed world) is to ensure that at least a modicum of the support work will go to local and regional operators if that is at all possible. Again, this refers to marketing – because it is incumbent upon the local and regional operator to convince clients that it has the distinctive capability to work alongside global operators and to play its particular part in the total development to the required quality, standard and deadline.

SUMMARY BOX 9.1 The Construction Industry and Other Global Developments

Four specific areas of activity where construction is certain to be required as the result of other developments can be identified:

- **Oil, gas and energy exploration and development projects:** as demand for energy grows, oil and gas fields hitherto considered uneconomic to exploit are now subject to serious scrutiny. There are large oil and gas resources in the former USSR, on the South American continental shelf and also in parts of South East Asia. For these to be fully exploited, both access, infrastructure and the means of distribution have to be built, as well as the facilities for exploration and exploitation of the resources themselves.
- **Transport:** as an example, the next generation of global public transport is envisaged to be a family of super-jumbo jets able to take between 500 and 1000 passengers. If these are produced, larger airport facilities are certain to be required in more parts of the world – and this includes take-off and landing areas, access infrastructure and airport buildings and facilities. Again, these projects are often commissioned with a view to wider economic development and the opening up of particular localities.

  It is also certain that road, personal and public transport infrastructure will lead to work for the construction industry. On the one hand, this is pioneering – developing infrastructure where no such thing exists; and refurbishment – where the infrastructure has existed for a very long while, but where sheer pressure of use and volume of traffic make it certain that it will have to be substantially upgraded. New materials – especially those capable of extrusion – are constantly being produced so that road infrastructure can be created more cheaply and last longer once it is in place.
- **Electricity, gas and water distribution:** again, this may be seen in two ways. First, in the developed (and relatively developed) economies and regions of the world, there is a demand for ever-higher quality of each, delivered at evermore cost-effective ways and methods. Second, in developing areas and those places where the population is burgeoning, it has not always been possible for the infrastructure development to keep pace. Especially in the latter, there is consequently plenty of work – for those companies that can present their expertise in these activities in ways demanded by the client.
- **Information:** the development of global information systems and satellite, fibre optic and personal transmission and receiving systems. Again, these have to be supported by a construction industry capable of building the facilities in which the points of access and distribution for these facilities are suitable.

Clearly, such projects require distinctive expertise and global scale resources. In many cases, these projects are simply too large for single organisations to contemplate. The rewards to be accrued from effective and successful activities in these spheres are enormous. It is also clearly possible to recognise that there is a prospect of organisations (as distinct from countries) creating their own trading blocks. The commitment to marketing in the pursuit of these rewards and scale of operations – and the long-term security brought once work is awarded – has to be seen in these terms. Effective marketing for this level of work is a major investment, requiring substantial commitment of expertise before any such work is gained; and requiring extensive client service during the period of the contract.

SUMMARY BOX 9.2 **Different Approaches to Contractor Globalisation and Internationalisation**

The following approaches may be illustrated.

The Eurotunnel model

The Eurotunnel model brought together a range of companies with distinctive expertise and also with large financial resources for the purpose of creating an enterprise that is both large and sophisticated enough to carry out the single project. In the particular case of Eurotunnel the organisation was created by ten building and civil engineering companies (five British and five French) and a financial resource base ultimately provided by 200 banks drawn from all over the world.

Because the company was created for a single project, the key feature of marketing activity is branding. It is the name of the company and identification with the given project that are of paramount importance. Initial marketing activity is therefore concerned with getting the name and project association as widely known as possible. Continuing marketing is concerned with building the desired image and quality of local relations, as well as commercial expertise in the minds of the community in which the project is being built, and to ensure that this reputation is built and enhanced. In the particular case of Eurotunnel, much of the responsibility for this in fact fell on the client; however, for other single-project activities carried out in this way, the problem may fall to contractor or client, or both.

The Japanese model

The Japanese model involves creating overseas and globalised subsidiaries in the image of the parent company; and of ensuring that all expertise that could possibly be required is kept under one roof. It involves high levels of investment and the creation of a relationship between the distinctive advantage of the expertise and services to be made with the needs and wants of the communities in which these are now to be made – the pressure is therefore on the single company to understand the specific markets of all activities, in all areas and of all its component parts.

The traditional multinational corporation
This has to be studied in the context that more and more organisations are trading outside the country of their own domicile in pursuit of new markets and competitive advantage, and also as they find alternative outlets for existing expertise.

From a marketing point of view, the problem here lies in familiarity. In the areas where they are well known, multinational marketing in the construction industry consists of expert client liaison and continuing levels of general familiarity. The same relationship and familiarity have to be engendered in the new areas. The attitude to marketing in new areas has therefore to be concerned with generating the same awareness and familiarity as exists in the familiar spheres of operation – and again, this means priority investment and levels of long-term commitment. The view should always be taken that, while clients and potential clients in new areas may have a general familiarity with the name and even the work of the multinational, a specific reputation has not yet been created. Attention, therefore, is necessary to the creation of this specific reputation in the new area.

The acquisition model
This is where the large organisation acquires either a local partner – either because it has to or because it wants to; or because it has chosen this way into the new market as the most cost-effective and client-effective form of operation; or because the organisation being acquired has distinctive expertise as well as reputation in the market that is to be served.

The marketing priority here is to ensure that the parent company and its new subsidiary can be presented as a strong and wholesome entity – and, again, this means spending time and resources on the familiarisation process, understanding and attending to specific needs and wants, and extensive personal and professional contact between the new organisation and its potential clients.

Joint ventures
These have been used extensively as a way of gaining a foothold in new markets. It is also a relatively safe way of testing the potential of a new market and understanding its drives, needs and wants; and also the true commercial potential (as distinct from that which is perceived).

Part of the commitment to a joint venture involves the use and harmonisation of subcontractor relationships in the particular situation; and of generating client liaison familiarity, understanding and confidence in the terms required in the new market. Again, therefore, marketing priorities have to be seen in these terms – and, again, this involves extensive human resource, expertise and commitment to presentation as well as delivery.

From this, a range of specific implications for marketing in the construction industry can be identified.

Competitive intensification

Competitive intensification is the key drive everywhere in the industry. As stated throughout the text, there are no safe or protected sectors. To all customers and clients, therefore, there is an ever-increasing range of choice available and rewards are dependent on the extent and quality of the investment that contractors are prepared to make in presenting themselves to best advantage in the areas in which they seek to operate.

This is transforming the priority order of contractor organisation activities. The nature of investment required, above all, is therefore concerned with:

- creating long-term, continuous and profitable contractor–client relationships;
- transforming service standards; and adopting the attitude that the delivery of construction projects is as much concerned with service as production expertise;
- building and creating much more comprehensive information and databases on all sectors.

Long-term relationships

The creation of long-term and enduring relationships with specific sectors and segments means investing in direct contacts, technological awareness, and product and service development. Creating this form of relationship means taking the following approach:

- gaining entry to client organisations and client bases through whichever doors are available: meeting and lobbying officials; public relations; client liaison staff so that a general awareness is built in.
- gaining access to professionals and experts in potential client organisations and government departments so that a real view of the likelihood of work can be gained and also the drives and restraints that cause contracts to be engaged;
- gaining access to other key players – politicians, officials, financiers – and establishing their needs, wants and drives and the basis on which they all would work;
- gaining access to other important and influential stakeholders and identifying potential pressure groups, lobbies and vested interests;
- assessing market volumes and duration, and the possible and likely share of this available to the particular contractor (see Summary Box 9.3);
- devising formats that meet everything to best advantage and that are acceptable in the local prevailing conditions so that distinctive expertise and commitment to service can be presented to best advantage also.

SUMMARY BOX 9.3 **Volumes, Duration and Potential**

At the macro end, a long-term and enlightened view is essential, For example, a £60 billion market that is likely to last for 50–70 years is certain to require investment in groundwork and research. It may also mean engaging in acquisition and joint ventures, and for overseas activities, engaging a local partner or front organisation.

The key issue lies in marketing to one's own stakeholders and backers. Shareholders and their representatives are used to short-term returns and dividends paid on annual profits. Other backers wish to see returns over 'a reasonable period of time' – and, again, this also means short to medium term. While this approach is designed to secure the eternal future of the organisation, staff nevertheless perceives that it is placed in a general state of flux and that there are certain to be changes – shake-outs, redundancies, redeployments – along the way. Even directors and top managers may not be inclined *per se* to take such a long-term, almost eternal view, of what is required.

The approach has to be seen in terms of the returns – and the investment is made in anticipation. The problem lies therefore in transforming the attitudes of the key players on the contractor side if the client base is to be convinced of the genuine seriousness of the approach and the determination to be a continued player in the sector.

This has finally to be seen in the context that (as we have seen earlier) the generally favourable response or positive reputation does not necessarily produce work. The nature and level of commitment necessary has therefore the prime purpose of getting over this barrier, ensuring that work is produced rather than a general perception that it might be.

Priority therefore lies in paying attention to the competitive requirements as much as distinctive expertise. Increasingly, competition is between organisations that can all do the work whatever it is, so the reasons for awarding contracts are based on the other factors.

Competitive emphasis

For all sectors in the industry, two distinctive points of view may be identified for the future – competing on price; and competing on factors other than price.

Competing on price

This is likely to remain a key feature of undertaking public service contracts in the UK and Western world. Moreover, as clients drive prices down, they too have increased expectations of quality, value and durability, and of extended professional contractor–client relationships. So the key concerns of those wishing to continue to play in these fields lie in:

- understanding this;
- structuring their cost, investment and expertise base so that it can be deployed profitably;
- maintaining the direct professional and political relationships necessary to ensure that work continues to be offered on a long-term basis.

For large private contractors, one of the drives is obviously on price. However, this is clearly much more than just one element in the mix of quality, quantity, value and time. Again, the relationship has to be serviced and maintained in this way.

Competing on factors other than price

This means attention to quantity, quality, value and time, and the other factors that go with these – product extension, pre and after-sales service; taking a more long-term view of contracts and the relationship potential, anticipating and identifying the range of possible needs – which may or may not require satisfaction; and identifying the key drives of the client.

This is also certain to mean acceptance and agreement of contract formats that have a greatly extended view of the contractor's obligations. In the house building and building products sectors, this implies accepting industrial standards and legally binding guarantees. At the macro end of the building and civil engineering industries, it means setting and presenting the highest possible levels and standards, and presenting these as an integral part of the contractor's product and service delivery.

Service marketing

Whichever the point of view, it follows from this that the marketing emphasis is certain to be on service (see Summary Box 9.4) and the commitment to deliver that which is promised, inferred or expected. The problem lies in understanding what the service commitment is from the contractor's point of view, and what the service requirements are from the client's – and this is only going to be achieved through creating enduring, direct, professional and personal relationships.

SUMMARY BOX 9.4 Service Marketing and Marketing Strategies

The following is a summary of the commitment necessary from given strategic points of view:

- **Cost leadership/cost advantage:** service advantage through attention to investment levels, prioritised and targeted marketing; and through attention to contractor staff expertise and understanding.
- **Focus:** service advantage through the ability to present specialised product and project ranges and expertise in accordance with differentiated (and perceived differentiated) client demands.
- **Differentiation:** the presence of a perceived quality advantage to those engaging a contractor; based on securing a perception of confidence, security and value in the minds of the key players in the client organisation.
- **Expansion, takeover, joint venture, local partner approaches:** based on the premise that product quality is already available in the sector, attention is paid to the mix of price, quality, quantity, volume and time – and quality, volume and time are direct service pressures while price is a reflection of a combination of product service.
- **First in field and pioneering:** product and project advantage initially; related to service and support if the market is developing.
- **Second in field/me too:** some products and project improvements; where the product is or becomes more or less generic, advantages in the service.
- **All-comers:** always service advantage.
- **New product development:** advantage is in product, quality and performance of new products; as more players come into the sector, advantage is sustained through product development and improvement, and through service.

Product and project extension

It follows from this increased attention to service that the 'product' of all the industries activities has to be extended – and this includes both the general nature of the package on offer, as well as specific attention to particular customer and client relations and requirements.

Creation of points of access and sale

Key features of product and service extension are convenience and access. The pressure is therefore on contracting organisations to commit themselves to both location and quality of access. This applies to enhancing the quality of access to current markets in response to the increased expectations of clients and also the activities of actual and potential entrants.

SUMMARY BOX 9.5 **Build and Operate**

One form of construction market development is the partial or full withdrawal of the client as the product or project end-user. The client's role becomes simply to agree and commission the work; and then to agree the format under which it is to be used.

Private Finance Initiative (PFI)
The UK government first created the PFI in 1989. Following intensive studies as to how financial resources could be profitably and effectively sought from the private sector in the creation of major public infrastructure and works necessary, a form of 'build and operate' was designed for this purpose. There are two main thrusts – agreeing to PFI type contracts puts the burden of project support from inception to completion on the contractor; payment for the work comes in the form of charges and levies on the end-users – and this may not cover the costs of the venture for very much longer than under normal contractor–client relationships (where the project is paid for fully on completion and delivery, and which may have been supported by interim payments along the way). The key to contract success lies in understanding the total range of commitment:

- to product and project quality;
- to end-user acceptability and service;
- to change of contractor activities from the design and build of facilities and projects to becoming an operator;
- to the creation of an enduring, profitable enterprise that may have asset value of its own;
- to the maintenance and upgrade of a facility as appropriate or desirable;
- to long-term broader acceptability, security and confidence.

Example: the Skye Bridge
This facility was commissioned in 1991 and completed in 1994 under the PFI. The contractor (Balfour Beatty) was given licence to operate the facility and to levy tolls upon completion.

The bridge replaced the ferry service that had operated between the island of Skye and the Scottish mainland for over a hundred years. The perceived advantage of the bridge lay purely in convenience – especially for tourists and other infrequent visitors.

The project ran into controversy following completion. The main objectors were the residents of the island, who felt that the charge levels raised were far too high – that the contractor was taking advantage of a captive market. Some people refused to pay; and those who did so were prosecuted. The contractor was also prosecuted for levying perceived unfair high charges.

The end result is an uneasiness throughout the industry concerning the range of possible outcomes of this form of activity while this is being decided by the Courts. Other contractors initially attracted by PFI type work are seeking forms of guarantees and indemnities against this form of prosecution before they will commit themselves fully.

It also applies in markets where entry is envisaged and represents a long-term commitment. For building, civil engineering, architecture, design, planning and consultancy services the outcome of this commitment is creating a familiar and enduring local reputation – and to do this, the organisation has to be perceived and known as local.

Some of this can be short-circuited to an extent through engaging in hook-ups, joint ventures and local partner arrangements, though even once these are achieved, the quality and response from the point of access have still to be guaranteed and generated. After all, the success of new entrants is based on some form of dissatisfaction with existing players, so the drive must be to harmonise the presence of the joint venturer or local partner with the distinctive expertise brought in by the new entrant.

The future of multinational corporations

Multinational corporations in the construction, civil engineering, expert services and architectural, planning and design sectors increasingly have to operate where the work is. The outcome of this is that in many cases the indigenous and familiar sectors become steady-state activities while the real competitive and profit advantages are to be found in the ability to penetrate and satisfy those areas of greatest growth. In simple terms, this means taking Western expertise, product and project quality and technological expertise to:

- the north – the former USSR and Eastern Europe;
- the south – Africa and South America;
- the east – China, Indonesia, the Philippines, Malaysia and Thailand – some of which are developed to greater extents than others;

and operating according to the norms, expectations and strictures of these markets. When put in these terms, there is not nearly enough expertise to go round. On the other hand, those who have gained dominant positions at present are seeking to protect these as far as they possibly can in order to secure their own long-term future.

In Western markets, these organisations have at present to recognise the state of the market as a combination of refurbishment and new project work, and attention to the quality of the built environment now demanded by clients, end-users and communities at large.

For the future, multinational operations are therefore certain to be based on the capability and willingness to understand and accommodate the opportunities and constraints indicated, supported by the willingness to undertake extensive, direct and first-hand market research and establish a long-term presence and commitment in this way.

Three distinct forms of multinational activity are apparent for the future – the Eurotunnel model (as indicated above); the complete entity (akin to the Japanese model indicated above); and the complete network.

The complete network

In this case, all that exists is a loose federation of networks of individuals and organisations and their expertise, which can nevertheless be deployed at short notice anywhere in the world. The success of this depends on the quality of the head office co-ordination function and marketing direction. The network is entirely dependent on the quality of the small central core, the initial point of contract – and the ability to engage the range of contractors necessary and to place them wherever they are required.

Indeed, in all cases the success of each depends on the quality of the head office, co-ordinating function and marketing direction. For the Eurotunnel model, this is the engagement of companies with distinctive expertise for the given purpose. The complete entity is dependent on continued market focus at head office – and in the case of many multinationals, this is likely to involve extensive head office reform.

Marketing and technology

There are three issues here:

- **invention:** the development of new products and project initiatives that are capable of improving and superseding the ways in which things are done and have been done in the past;
- **capability and expertise:** the ability to implement new ways of doing things;
- **willingness:** above all, the willingness to engage expertise in ways demanded by new markets. This also refers to the willingness to accept and embrace new ways of doing things, especially where these do not accord with the industry's current and historic norms and ways of working.

Some of this refers in part to previous chapters. It is no use having the best quality architects, designers and planners if this is not recognised by the markets or if the particular practices are not prepared to deliver their expertise in the ways demanded by clients – and this has to include speed of response and relationship development as well as professional capability. It is no use inventing plastic-based extruded road and highway materials if the benefits of this cannot be presented to the particular client base – as well as product durability, other benefits such as speed of completion, value, refurbishment and upgrade have to be considered.

High levels of expertise, technology and capability have therefore to be seen in terms of customer demand and satisfaction. The emphasis is shifted from the provision of product quality and professional expertise to the capability to use these to satisfy client demand. And, again, where politicians and other non-specialists are the commissioners of work, quality, value and expertise have to be presented – marketed in terms that they understand and to which they will respond positively.

Small and medium sized companies

On the one hand, there is a good strong history of successful and effective marketing in some small and medium sized companies, especially in the regional and local house building and smaller scale building projects, double glazing and civil engineering subcontractor sectors. In these cases, marketing has been based on a combination of personal relationships and reputation, supported by known, understood attention to specific client demands. Where necessary – especially in double glazing and house building – more general advertising, presentation and promotional features have been included, with varying degrees of success, especially in times of relative prosperity (e.g. the UK housing boom of the late 1980s), or when pitching at relatively prosperous sectors (e.g. secure and sheltered accommodation for medium to high income brackets).

On the other hand, this has to be seen in the context that these sectors are also becoming evermore competitive in their own way. More players are competing for the same markets and this has led to:

- downward pressure on prices;
- upward pressure on quality, reliability and durability;
- pressures to extend the quality and nature of contractual relationships – especially in the post-completion phase where after-sales commitments are increasingly required and expected.

Moreover, larger players are extending their activities downwards into smaller markets. Just as multinationals are taking (for them) smaller jobs in their own domestic markets, regional and medium sized players are taking on work hitherto carried out by small local companies. The result is a greater competitive pressure all round, and again, therefore, increased competitiveness – and the need for marketing expertise.

Professional services

The same also applies to technological expertise, architecture, planning, design and professional consultancy services. Until very recently, these organisations existed on a combination of reputation and access based on steady and assured client bases and known (or almost known) volumes of work.

This has changed; larger consultancies and practices have gained market entry and share at the expense of small organisations. Moreover, once they have gained entry into particular niches, they have been able to compete on knowledge and understanding, presence, price and deliverability – exactly the qualities of the previous players. They have therefore used niche and entry opportunities as key features of their marketing. The result is to force smaller, local and specialised practices to rethink their own position entirely – and many have been forced out of business. Those remaining have had to regain their position either by repositioning themselves entirely, or by rejuvenating the marketing efforts already made. For the client base, there is no particular advantage in dealing with an international or multinational company – but it will do so if it is made to feel that it is in its interests.

MARKETING AS A PRIORITY

From whatever point of view, the key to an effective and profitable future lies in the ability to develop, in turn, a professional and expert marketing approach along the lines indicated throughout the text. The context in which this is carried out is the combination of the circumstances indicated above. The specific mix of these clearly varies between organisations and between the markets served (and in which it is intended to operate). The following factors are covered:

- Marketing objectives: in terms of product, price, place, promotion, expertise, capability, willingness, quantity, quality, value and time.
- Situational analysis: general scanning; market audits and research; market analysis; customer and consumer analysis; end-user analysis; client analysis; competitor analysis; and internal capability analysis.
- Assumptions for the present and future: environment and location; markets; customers, consumers and end-users; clients and potential clients; and competitors. Testing these assumptions and researching them; getting over the perceptual barriers; gathering market intelligence; classifying and codifying information.
- Defining critical success factors and using these as benchmarks for progress.
- Identifying a strategic position from which the markets and potential markets are to be approached and served.
- Taking short, medium and long-term views of activities; understanding the levels of investment necessary to achieve these; prioritising

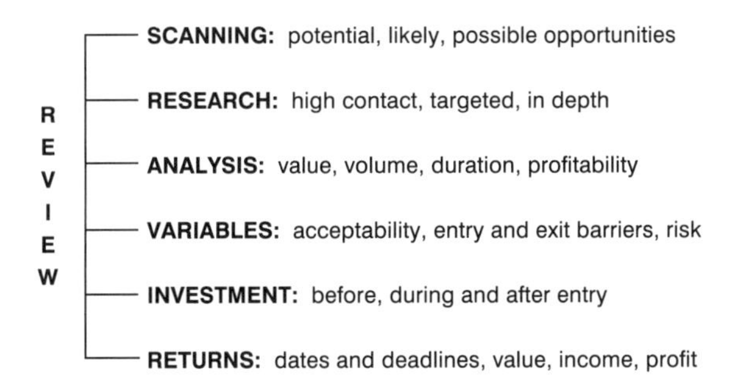

Figure 9.1 Marketing planning for the construction industry.

marketing activities accordingly; engaging the necessary expertise and capability.

- Establishing marketing mixes for each client, client base, locality and region (including the domestic).
- Identification of potential problems and blockages; taking the best/medium/worst outcomes approach to each activity and client.
- Authorisation of marketing activities.

See Figure 9.1 for a summary of marketing planning.

A note on implementation and review

At the implementation stage, the following points are important:

- Results are judged against expectations: expectations must therefore be carefully assessed in advance; and if there are misperceptions of these, these should be handled at the inception stage. Of specific importance are the expectations of stakeholders and backers, to which reference has already been made.
- Instant success is not a guarantee of long-term or future success and, again, this has to be seen in the context of what the successes actually are.
- Instant failure (or perceived failure) is not a guarantee of long-term and continued failure; this must be assessed in the light of circumstances, events, expectations and projections. Above all, where the stakes are very high and where the rewards for gaining access to a particular market are steady, long-term, assured quality and volume of work, instant failure (or perceived failure) is likely to require an evaluation of activities to date rather than the need to withdraw as quickly as possible.
- Where failure to gain market share or foothold is apparent, with-

drawal may be essential. However, it has to be seen in the context that if not handled properly, it may cause damage to reputation and confidence in those markets hitherto served effectively.

- The quality of review is dependent on the capabilities and expertise of those undertaking the review and their own particular perspectives and standpoints. Again, this refers to particular stakeholders – especially those who envisage short-term returns. It also refers to those within the organisation not directly concerned with marketing and who have their own particular pressures and drives. For example, the difference between investment and cost accrual is very much a matter of perception – one person's investment is another's pay out.
- Blockages and barriers have to be seen in terms of: the extent to which they are key or critical; the problems and advantages incurred in overcoming them; their general extent and prevalence – whether the rewards to be gained are worth taking time, energy, commitment and resources to tackle the problems.
- The location of decision-making is also important. This also refers to the extent to which marketing takes key marketing decisions or whether these are nevertheless taken by those who have no direct marketing expertise.
- Internal factor assessment, especially that concerned with matching contractor expertise to client expectation – have to be assessed as marketing activities go along. As we have seen in many cases, this is likely to be as much a matter of willingness as capability.
- External factor assessment is also essential. As we have seen, this is also certain to include attention to the priorities of pressure groups, lobbies and vested interests as well as client professional demands.

Marketing in all aspects of the construction industry is therefore concerned with mixing high levels of professional expertise and capability with the willingness to present these to best advantage in terms of particular clients and client bases. Indeed, the main problem that has to be addressed throughout the industry is concerned with raising the level of understanding of what marketing is and how its principles (such as they are) can be best and most effectively applied, so that what can be done is known and understood by clients. Once this attitude is engaged, the industry will have taken giant steps towards becoming truly competitive; and those companies and organisations that do not see marketing as a priority will be able to compete for work on the same basis as those that do. The increased professionalisation of marketing and the adoption of its principles and practices are the most critical factors facing the construction industry as it prepares to secure its long-term future.

In summary, the implications for all construction professionals are as follows:

- the need to place client need rather than contractor expertise at the core of all activities;
- attention to all aspects of project design and inception, and the development of presales services and activities, the service aspects of work in progress, and after-sales activities (including long-term user management, facilities and environment management, as well as the maintenance of the finished project itself);
- attention to the quality and durability of the project when complete;
- acknowledgement of the wider interests, needs and demands of all parties concerned – whether acting in the contractor or client interest, or as an independent expert;
- acknowledgement that everything can be improved – and this includes all aspects of construction provision including materials, materials usage, expertise, logistics, planning, design, technology and calculation;
- acknowledgement that presentation is as important as technological expertise;
- acknowledgement that there is now no such thing as a closed or protected market, and that every construction market is open to global competition;
- acknowledgement that as the industry is globalised, this means having global knowledge and understanding as well as a global presence – indeed, the global presence is no use without knowledge and understanding, and this includes learning and understanding the cultural and presentational aspects that are important to work commissioners in different parts of the world;
- continuous attention to professional, technological, expertise and service development – failure to do this will necessarily mean loss of competitive advantage to those companies that do so;
- continuous attention to service, service extension and service development in all aspects of the construction process – and, again, whether working for contractor, client or as an independent expert;
- acknowledging the full range of responsibilities inherent in carrying out work, wherever this may be. This means taking a truly enlightened view of the needs, wants and demands of clients and commissioners of work; the pressures on contractors and consultants; and understanding the real, total and enduring impact of activities on the particular environment and location in which these take place.

Bibliography and References

Adams I., Hamil S. and Carruthers A. (1991) *Changing Corporate Value*, Prentice Hall, Englewood Cliffs, New Jersey.
Ansoff H.I. (1979) *Strategic Management*, Pelican, London.
Ansoff H.I. (1984) *Business Strategy*, Pelican, London.
Ansoff H.I. (1986) *Corporate Strategy*, Pelican, London.
Asch D. and Bowman C. (eds.) (1986) *Readings in Strategic Management*, Macmillan Press, Basingstoke.
Ash M.K. (1984) *On People Management*, Warner, New York.
Baker M. (1983) *Market Development*, Penguin, London.
Boston Consulting Group (1970) *The Product Portfolio in Concept*, BCG.
Bowman C. and Asch D. (eds.) (1992) *Strategic Management*, Macmillan Press, Basingstoke.
Buell V. (1985) *New Product Development*, Sage, London.
Buzzell R. and Gale B. (1987) *The PIMS Principles*, Free Press.
Chattell A. (1995) *Managing for the Future*, Macmillan Press, Basingstoke.
Christensen C.R. (1986) *Business Policy*, Irwin.
Clarke W. (1988) *The Want Makers*, Corgi, London.
Cohen G. (1988) *Marketing Strategies*, Routledge, London.
Doyle P. (1989) *The Strategic Importance of Brands*, Economist Intelligence Unit.
Drucker P. (1964) *Managing for Results*, Heinemann, Oxford.
Drucker P. (1985) *Innovation and Entrepreneurship*, Pan, London.
Drucker P. (1981) *Towards the Next Economics*, Heinemann, Oxford.
Hamel G. and Prahalad C. (1994) *Competing for the Future*, Free Press.
Johnson G. and Scholes K. (1994) *Exploring Corporate Strategy*, Prentice Hall, Englewood Cliffs, New Jersey.
Kanter R.M. (1986) *The Change Masters*, Free Press.
Kotler P. (1980) *Marketing Management*, Prentice Hall, Englewood Cliffs, New Jersey.
Kotler P. (1993) *Marketing Management: Analysis, Planning and Control*, Prentice Hall, Englewood Cliffs, New Jersey.
Macray C. (1991) *World Class Brands*, Addison-Wesley, Reading, Massachusetts.
Packard V. (1957) *The Hidden Persuaders*, Penguin, London.
Packard V. (1960) *The Waste Makers*, Pelican, London.
Pettinger R. (1995) *Introduction to Corporate Strategy*, Macmillan Press, Basingstoke.
Pettinger R. (1997) *Introduction to Management* (2nd edn), Macmillan Press, Basingstoke.
Pettinger R. and Frith R. (1996) *Measuring Business and Managerial Performance*, STC.
Peters T. (1990) *Thriving on Chaos*, Macmillan Press, Basingstoke.
Peters T. (1992) *Liberation Management*, Macmillan Press, Basingstoke.
Peters T. and Waterman R.H. (1982) *In Search of Excellence*, Free Press.
Porter M.E. (1981) *Competitive Strategy*, Free Press.

Porter M.E. (1986) *Competitive Advantage*, Free Press.
Porter M.E. (1990) *The Competitive Advantage of Nations*, Free Press.
Randall G. (1992) *Principles of Marketing*, Routledge, London.
Sternberg E. (1994) *Just Business*, Warner, New York.
Toffler A. (1980) *The Third Wave*, Collins, London.
Ward J. (1984) *Profitable Product Management*, Heinemann, Oxford.
Webster F.E. (1991) *Industrial Marketing Strategy*, Prentice Hall, Englewood Cliffs, New Jersey.
Wickens P. (1995) *The Ascendant Organisation*, Macmillan Press, Basingstoke.

Index